LOSING GROUND

LOSING GROUND

Identity and Land Loss in Coastal Louisiana

David M. Burley

Foreword by Sara Crosby
Afterword by T. Mayheart Dardar and Thomas Dardar

University Press of Mississippi
Jackson

www.upress.state.ms.us

The University Press of Mississippi is a member of the Association of American University Presses.

Frontis photograph courtesy Susannah Bridges

Copyright © 2010 by University Press of Mississippi
All rights reserved
Manufactured in the United States of America

First printing 2010
∞
Library of Congress Cataloging-in-Publication Data

Burley, David M.
  Losing ground : identity and land loss in coastal Louisiana / David M. Burley ; foreword by Sara Crosby ; afterword by T. Mayheart Dardar and Thomas Dardar.
    p. cm.
  Includes bibliographical references and index.
  ISBN 978-1-60473-488-1 (cloth : alk. paper) 1. Coastal zone management—Louisiana. 2. Coastal settlements—Louisiana—Psychological aspects. 3. Coast changes—Louisiana—Psychological aspects. 4. Gulf Coast Region (La.)—Population—Environmental aspects. 5. Place attachment—Louisiana—Gulf Coast Region. 6. Identity (Psychology)—Louisiana—Gulf Coast Region. I. Title.
  HT393.L6B87 2010
  304.2'50976309146—dc22   2009046390

British Library Cataloging-in-Publication Data available

For Susi and Max . . . and Harvey and Sophie! Thank you for all the joy.

CONTENTS

Foreword ix

Acknowledgments xiii

Chapter One
Introduction 3

Chapter Two
Losing Louisiana: A Meditation on Coastal Land Loss 15

Chapter Three
Communal Histories 27

Chapter Four
This Is Our Home: Attachment to the Coast 39

Chapter Five
Seeing It for Themselves 59

Chapter Six
Saving Place: Residents and Their Environment   110

Afterword: The Path Ahead   132

Appendix   136

Notes   153

References   159

Index   164

FOREWORD

My grandma wrote a book about Grand Isle, the only inhabited island off the coast of Louisiana and our family home for over two hundred years. She called her book *Seven Miles of Sand and Sin* and kept it close beside her living room chair in a Square Deal note pad with "39¢" penciled on the cover. Fortunately for the reputations of our neighbors, she never quite got to the sin part, but she does describe the island as a "paradise . . . plentiful of all God's beauty and his gifts." In one of my favorite passages, she muses that "[l]ong before I was born there were large orange groves here and all the boats had sails. When it was spring time and the boats carrying supplies to and from our island got lost in the fog they could tell when they neared the island by the smell of orangeblossoms." She breaks off this sweet reverie, however, with a disappearance: "Now there isnt an orange tree left. The hurricanes and dease [sic] killed them all." Of course, island residents replanted or tried to. (Even my own parents babied an orange sapling, but some trifling surge from the gulf stifled it before it could struggle past adolescence.) We continued to cultivate a hardier "paradise." Grandma talked about pomegranates, plums, and grapefruit, and I grew up relishing mulberries, figs, peaches, and sharp island bananas. But they've disappeared, too, now. Hurricanes and saltwater intrusion have poisoned all but a few grim little guava bushes.

In *Losing Ground: Identity and Land Loss in Coastal Louisiana*, David Burley aptly describes the process that killed the island's fruit trees as a "slow disaster." Unlike most "natural disasters," however, this one is not unavoidable, inevitable, or even natural. It is the result of an ongoing human decision.

For eons, in the Mississippi's slow struggle with the Gulf of Mexico, the river trounced the sea, and each spring rush poured black sediment and freshwater into the delta, building up the coast and barrier islands and nourishing its marshes and bayous. But in the last fifty years, south Louisianans have been living with a creeping Frankenstein of a catastrophe caused by dredging the Mississippi, putting up levees along its banks, sucking petroleum from under the land, and slicing canals through the marsh. These activities facilitated urban development and the extraction of oil and natural gas from Louisiana, but, by dumping the Mississippi's annual tribute off the continental shelf and opening a dehydrated delta to the gulf's salt waters, they have reversed the traditional outcome of that contest between river and sea. The results have been devastating. First, and most obviously, south Louisiana has suffered and continues to suffer monumental land loss, which includes the marshes and barrier islands, like my home, which used to absorb hurricanes before they reached population centers such as New Orleans. Second, this erosion is destroying the single most intensely productive ecosystem on the planet. For instance, coastal Louisiana nurses into life almost half the seafood consumed in this country (as well as untold numbers that got away) and provides sanctuary to a similar proportion of our migrating songbirds. I remember sitting on the porch with Grandma and watching the sky flash into a mass of crazy colors, as hordes of half-starved songbirds dove into our mulberry trees. It was their last chance to fortify themselves for the grueling passage to South America. I know that many of these tiny lives didn't survive even then, when they had profusions of purple berries to sustain them, and I wonder how many more now struggle up from dead trees to drop into the gulf.

Other books, however, have documented the sufferings of songbirds and widespread environmental impact of coastal erosion. *Losing Ground* instead tackles a more intangible issue: how this devastation affects the region's human inhabitants. Personally, I've seen "erosion" form into a collection of mysterious and shamed absences: not only the missing birds and trees and land but language. It's a strange still oppression, tincturing ordinary gossip with something between horror and embarrassment. I've seen its operations so many times. When people talked about "the erosion" at the supermarket or on porches, everyone put on the brave face—a shrug, a joke, the laugh that's just a bit too loud, then a swift hush and the look away. I don't think those haunted faces reflect only "practical" concerns about lost property or livelihood. I think for a split second we all wonder, "What happens to the dead ones we loved when the land disappears?" Other parts of America erect statues and put up historical markers to record epic heroes and battles and remind themselves of who they were and where they've been, but, well, we're a bit too quiet for epic and a bit too poor for monument. We instead inscribe our stories into the landscape, and the narratives are rare and personal and strange. (Perhaps a bit too often, they memorialize the "sin" that Grandma left out of her book.) We've remembered who we were because we never moved. My family has lived in the same cypress house for two hundred years, and most of our neighbors trace their generations in south Louisiana at least that far. These generations farmed, fished, withstood brutal hurricanes, and sometimes pirated when Jean Lafitte was about. They've stayed with us in the shared tombs and homes and land, and now watching its dissolution provokes at best a sense of helplessness and at worst a sickening suspicion of our complicity. We turn away from each other out of shame and fear that we might just see that treason mirrored and confirmed in our neighbor's eyes.

At the same time, this embarrassment has kept us from looking to our fellow Americans. A massive public work like restoring coastal Louisiana requires national will, but it's so shameful to be a victim, especially a victim of "nature"—whether of God or God in the market. This slow disaster combines both. There's also something effeminate about an enduring commitment to place, something morally suspect in a country that has traditionally idealized rootless, self-sufficient self-reliance that has lately blended with perverse selfishness. We're afraid to plead and hear, "Why don't you just move?"—a statement which implies other judgments like, "What's wrong with you? Do you want to be a victim? Don't you have any get-up-and-go?"

But where can you go? And for how long?

I moved to Ohio. I like Ohio. I'm happy here, free of that hovering oppression, mostly. But I don't ever want it to descend upon my daughter, and the best scientific evidence suggests that "slow disaster" is creeping into Columbus, too. Perhaps the ground will not literally disappear from under her feet, but we are entering a time of accelerating global climate change, when loss will mark if not our own lives then our children's. And *Losing Ground* asks how we—the people of America—plan to deal with it: "[H]ow much are we willing to sacrifice as 'not important enough'?" If we decide that south Louisiana—despite its crucial ecological and economic contributions—is "not important enough" to save, then what next? Or, again in Burley's words, "As we start to see the effects of climate change in coastal Louisiana, elsewhere in the nation and around the globe, are we willing to let our environments, the places we live or don't live, become irreparably damaged?" Ultimately, "Are we all expendable?"

Coastal Louisiana, Burley suggests, can serve either as a model for our nation's future success or an exemplar of its degradation. His work has revealed coastal residents desperate to do the hard work necessary to restore our southern border and create a sustainable environment. He argues that we need "participatory research" that meshes together local knowledge and energy with outside expertise and resources. We need a nation dedicated to empowering rather than shaming each other.

In short, we have a fight on our hands, and we can establish the front on the Gulf Coast or later and to our greater peril in the heartland. To echo Hillel the Elder, "If not here, where? If not now, when?" Perhaps I don't need "paradise," but, without it, I just might lose Ohio.

Sara Crosby
Assistant Professor of English
The Ohio State University at Marion

ACKNOWLEDGMENTS

I would first like to lovingly thank my wife, Susannah Bridges, and my son, Max Gisleson. Susannah, with her advice, artistic eye (book cover and pictures selected for the book), passion for southern Louisiana and all of life, and playful exuberance, has been a source of continuous support. Likewise, Max continues to teach me about life and myself. His generous and humorous spirit is a source of inspiration and joy.

Next, I owe much gratitude to Dr. Pamela Jenkins and Dr. Shirley Laska, without whom the research for this book would not have been possible. Dr. Jenkins has served as my teacher and mentor and has inspired me with her passion for qualitative research and the desire to serve communities. She has provided me with endless counsel throughout my academic career, and she critiqued the early forms of this work well. She pushed me to do my best. In addition, Dr. Jenkins was the catapult that launched me into this research. She recommended me for a research assistant position at the Center for Hazards Assessment, Response and Technology (CHART) at the University of New Orleans where I met another mentor, Dr. Shirley Laska. Dr. Laska, founder of CHART, a multidisciplinary disaster research center stemming from the sociology department at the University of New Orleans, held great confidence in me and urged me to continue the center's first project which they initiated, the Coastal Communities Project, of which the work contained here is an extension. Dr. Laska's passion for coastal Louisiana and pursuit of funding allowed this project to bloom, and her guidance, along with the direction of Dr. Jenkins, and trust gave me the freedom to conduct aspects of the research for this project, especially the field research, in a manner that developed tremendous professional, academic, and personal growth. Truly, the field research for the Coastal Communities Project was one of the most memorable experiences of my life. Their public sociology work continues to inspire as they tirelessly research, advocate, and engage life in post-Katrina southern Louisiana. One of those who tirelessly advocated for coastal restoration, the departed coastal geologist Dr. Shea Penland, director of the Pontchartrain Institute for Environmental Sciences, deserves thanks from everyone concerned about coastal Louisiana. He is missed.

Dr. Ivor van Heerden is another who advocates for coastal Louisiana. I thank him for his research and for speaking out at great risk to his job about the bureaucratic failures associated with coastal land loss and Hurricane Katrina. I would also like to thank Dr. David Gladstone, Dr. Robert Whelan, and Dr. Steve Kroll-Smith for their help with the early forms of this work. Dr. Gladstone and Dr. Whelan were extraordinary teachers and always open for counsel. Dr. Kroll-Smith was also the press reader for this manuscript. His suggestions were invaluable. He has been a mentor for many years and continues to be an invaluable and magnanimous guide. His constructive criticism always leaves colleagues and students feeling positive about their work.

I would also like to thank all the people who worked on the Coastal Communities Project. Friend Dr. Joanne Darlington DeRouen and I walked, sometimes blindly, through the first stages of field research in Grand Isle, Louisiana. Brian Azcona, colleague and friend, conducted rigorous historical and other content analysis while Traber Davis, also colleague and friend, provided almost every kind of support, technical and organizational, and she served as collaborative coder for the data to ensure reliability of findings in analysis. In addition, she was a wonderful field research partner for the community of Lake Catherine and those in south Terrebonne Parish where she urged me to submit a photo essay to the American Sociological Association's journal, *Contexts*, our first publication, yielding great satisfaction at bringing the communities we were researching to the attention of a mass public of scholars and students. Many of those pictures are reprinted here. To all those who worked on the Coastal Communities Project and with whom I worked at CHART I am grateful. Also deserving of my gratitude are my editor, Craig Gill and everyone at the University Press of Mississippi, Anne Williams, and Tim Carruthers. Craig has been open and cooperative from the beginning, and Dr. Carruthers's photographs of coastal Louisiana for the University of Maryland Center for Environmental Science are illuminating. Thanks to Anne Williams, a good friend, who likes to play with computers and adjusted the map of the coastal communities for this book. Finally, I would like to thank my loving parents, Iris and Irving, as well as Nona and Big Daddy, who always take an interest and are supportive.

Further thanks go to Dr. Sara Crosby, a native of Grand Isle, and Thomas Dardar, a native of south Terrebonne. Sara's foreword for this book cuts to the experience of coastal land loss in southeastern Louisiana with the artistry that only a writer of literature can deliver. I am emotionally moved each time I read it. Thomas Dardar, a councilman for the Houma Indian Tribe, wrote the fabulous afterword. He has been generous with his ongoing conversations and meetings with me. His principled and pragmatic outlook fuels his constant advocacy for his community, both conventional and unconventional. Finally, I thank the people of southeastern Louisiana. We are they and they us.

LOSING GROUND

1

AN INTRODUCTION[1]

In the early winter of 2003, Cheyenne, a woman in her early fifties, told a story about her home.

> When you build a house, you expect your house to last fifty years. You are going to pass your house down to your kids. But if they don't do something about the erosion, this will not be here in fifty years. Because that was a field where my husband used to play right across that cement road, and it's marshland right now. If it wouldn't be for that cement road, my yard would be marsh. They have a certain type of grass that grows in the marsh, it doesn't grow in people's yards, and now I have it in my yard. So that's saying my yard is marshland. It's kind of depressing. And when we built our house, you had to build twelve and a half feet above sea level. Now you have to build fourteen feet above sea level. The houses are so high now. The water table is coming up. And we don't know if the water table is coming up or the land is sinking or both. But for some reason every storm, the water gets higher. Because that was the most water I have ever seen after these last two storms. And these storms didn't even hit us directly. When I built my house twenty years ago, I would have never thought that there's a chance that it's not going to be here. Not that it's going to disappear, but if they don't do something about the erosion, it's going to be just water. Because that's water right there. That used to be hard land. They had pecan trees, they tell me, when my mother-in-law was a kid, like forty-five

years ago. She said there was a big pecan grove. It's swamp right now. That's all the proof I need that fifty acres a year are going off in erosion.

Cheyenne owned a small seafood distribution business along with her husband in St. Bernard Parish (in Louisiana parishes are like counties in the rest of the U.S.). They lived in coastal Louisiana, but it is not as if they lived on or near a beach. There is no cliff or real shoreline. You can't see the Gulf of Mexico from where Cheyenne lived. Her home and community were on hard land surrounded by cypress treed forests and bayous that look like canals or small rivers that meander into small lakes and eventually wind their way to the gulf. Cheyenne's parish of St. Bernard juts out in an easterly direction into the Gulf of Mexico and is adjacent to and southeast of Orleans Parish where the city of New Orleans lies. Cheyenne and her husband are both natives, descendants of Spanish settlers who called themselves Isleños and were some of the first Europeans to settle the area beginning in 1778. Indeed, the parish has a long social history. Archeological findings of Native American complexes date back more than two thousand years (LSU AgCenter 1998).

But now this place is fading fast. Cheyenne spoke these words after facing Hurricane Lili and Tropical Storm Isidore during the fall of 2002. Even though her community only caught the edges of those relatively weak storms, she said, "that was the most water I have ever seen." Never before had storms of that magnitude brought so much water. The loss of land is no sudden mystery to Cheyenne. She, along with the rest of her coastal Louisiana neighbors, had been watching the land slowly disappear into open water for some time.

Her urgency was apparent from the beginning of my interview with her. We sat down in her kitchen; I explained the interview process and got the mikes and digital recorder set up. As I asked the first question about what it was like to grow up in coastal Louisiana, she sprang up from her chair, walked briskly through her living room and out to her front porch while I stumbled in tow trying to take great care not to interrupt the recording. We stood on the porch twelve feet above ground as she pointed at her front yard, the land across the street, and the road, which was a good two to three feet above these two sections of land. She launched into the above passage, explaining how different and, more important, solid and lush this all used to look. She pointed out that, across the street, just beyond a dying line of cypress trees, the bayou was growing wider and higher. The water was creeping closer.

In August 2005 her community was totally lost to Hurricane Katrina. They are still rebuilding, in another location in St. Bernard Parish.

Coastal Louisiana *is* vanishing. Hurricanes Katrina and Rita only exacerbated an ongoing eradication of the region. The land has been disappearing for generations. Some of it is natural, but the rate of land loss exploded in the 1950s as two major factors coalesced: the leveeing of the Mississippi River and oil and gas activity in the region (LCA 2004). Levees along the Mississippi River salvaged communities from annual flooding and promoted more development. However, those levees cut off the deposition of sediment in the region. Southeastern Louisiana was built by a yearly gift of soil and sediment, a donation from the rest of the country carried by the tributaries of the Mississippi River and ultimately down the great river and into southern Louisiana. No longer receiving annual deposits of sediment, southern Louisiana began to sink and become more vulnerable to storms which wash away land and increase saltwater intrusion from the Gulf of Mexico.

Oil and gas activity is the other major factor that expedited Louisiana's land loss in the second half of the twentieth century. While welcomed into southern Louisiana as a sort of economic savior, the oil and gas industry also produced negative consequences. The industry cut thousands of miles of canals into the coastal region, which allowed salt water from the Gulf of Mexico to permeate the region, killing the vegetation that prospered in a brackish environment. Since the 1950s Louisiana has been losing an average of about thirty-four square miles of land annually (LA Fact Sheet 2004). Among the consequences of this land loss is the increasing vulnerability of Louisiana to storms. The storms of late summer 2005 effectively exposed that vulnerability.

Hurricane Katrina arrived in southeastern Louisiana on Monday, August 29, 2005, and a few weeks later, on September 24, Hurricane Rita made landfall along the Louisiana and Texas border. The dire predictions from hundreds of academics, scientists, and state and local officials had come to fruition. As bad as the devastation was, it could have been much worse. In particular, if Katrina had traveled a few miles farther west, the eyewall and the most destructive segment of the hurricane (the eastern portion) would have scoured New Orleans much more brutally. Indeed, it is hard to fathom.

But Katrina and Rita had a much more powerful impact than they would have had in the mid-twentieth century. In decades past, similar hurricanes would have been weakened by a robust set of barrier islands and wetlands. Louisiana's wetland system, which comprised as much as 40 percent of the nation's wetlands during the middle of the twentieth century (Louisiana Department of Natural Resources 2004), weakens storms as they move inland and, thus, serve as a buffer for New Orleans as well as smaller towns and communities that dot the coastal region. However, since the 1930s Louisiana has lost over nineteen hundred square miles of wetlands (LCA Fact Sheet 2004). Consequently, with less and less land to weaken them, storms like Katrina, Rita, and, more recently, Gustav and Ike in September of 2008, reap much more destruction than they would have in, say, 1965 when Betsy, the last "big storm" to hit New Orleans, made landfall.

A result of the damage incurred by these storms, especially that of Katrina and its subsequent torrent of media coverage, was that Louisiana's coastal land loss was brought to the attention of the nation's citizens and leaders. Large-scale restoration proposals once thought too big to gain adequate funding seemed to gain new life. However, even though the hurricanes of 2005 gave necessary notice to the decimation of Louisiana's coast, restoration remains uncertain. Restoring the region's coast involves rebuilding a vital ecosystem damaged by natural processes and accelerated by human encroachment. True, some coastal erosion is natural, the causes of which are from wave action and storms, but these natural losses are exaggerated and quickened by human activities. For decades, the communities of south Louisiana, along with local scientists, have been clamoring for restoration support. As a result, there are some restoration projects in place, but these are relatively small-scale and piecemeal. More recently, the state has joined the chorus to obtain region-wide funding by mounting a national media campaign and lobbying effort dubbed "America's Wetland: Campaign to Save Coastal Louisiana" (America's Wetland 2006).

Prior to Katrina and Rita, it seemed that the residents of Louisiana's coastal communities were skeptical of government and the possibility of any significant restoration. Now, it is certainly hoped by many that the devastation wrought by these storms will bring weight to the urgency and importance of Louisiana's coastal lands and compel the federal government to dedicate the needed resources to its restoration.

Indeed, the massive loss of land is something that concerns all of us. Economically the region holds great importance in terms of the nation's seafood industry, oil and gas infrastructure, and international transportation of goods and resources. Also, the migration of people that would result from the extinction of communities would put tremendous strain on surrounding cities' services and infrastructure as well as take major national financial resources to ease that strain. Additionally, we might have to ask ourselves how much are we willing to sacrifice as "not important enough"? As we start to see the effects of climate change in coastal Louisiana, elsewhere in the nation, and around the globe, are we willing to let our environments, the places we live or don't live, become irreparably damaged? What does that say about us, how much we value ourselves and each other?

Coastal Louisianians are hoping the rest of us see how important they are. In what is commonly referred to as "the post-Katrina era," they have more reason than before Katrina and Rita to believe that the restoration of this vital place might occur. Yet, little has happened thus far. Sobering experience has caused the skepticism that was present prior to the storms to carry over into its aftermath. So, coastal residents are probably crossing their fingers but are far from overconfident.

Part of this cautious outlook is the fact that most residents of coastal Louisiana and all of the communities included in this study were substantially impacted by Hurricane Katrina and/or Rita. Some communities—Lake Catherine, those of lower Plaquemines and eastern St. Bernard parishes—suffered total devastation. Many residents from these communities remain displaced, their homes and communities completely destroyed. The other communities—Delcambre, Grand Isle, and those of south Terrebonne Parish—endured widespread damage but did not suffer total annihilation, and while they are a little further along in the rebuilding process than the former communities, they are still in the midst of this slow process. The residents of these communities have a long history along Louisiana's coast. They are deeply attached to this place and have been calling for action while watching the erosion of their landscape.

THE CALL BEFORE THE STORM

This book asks, "What is it like to experience the loss of place?" It addresses the significant issue of human identity and environment. This topic, one that is taken for granted in most of our daily lives, is situated within the context of acute and chronic loss of viable land. It answers the question "What is it like?" by probing into what Louisiana's coastal residents said about watching the places where they live disappear. The voices of the residents sent out a plea to alleviate the damage, to stop the bleeding. Even at the time the interviews were conducted, 2002–2004, they realized they could not do it on their own; it was too widespread. They spoke with an air of urgency that exemplifies a fear of losing not just the physicality of place, but their identity as well. While analyzing the data, residents' voices, it was evident that they were calling for assistance. Yet now, in light of the storms, their call seems prophetic. As outlined in the subsequent chapters, the meanings that residents attribute to coastal land loss reflect their sense of self.[2] And their sense of self reflects their identification with place. In other words, residents come to see themselves through the landscape, consisting of both its physical and social elements. Consequently, place comprises a significant part of identity. Because place is such a substantial part of these people's sense of who they are, processes that occur in the landscape, such as coastal land loss, impact their identity.

At the onset of this study, it wasn't obvious that place, or more specifically, people's attachment to place, would instruct how residents understood land loss. It was thought that residents might look at the issue in relation to how it affected their occupation or community more than that they would interpret the event through an overarching attachment to place.

I came to this work due to my involvement with CHART, the Center for Hazards Assessment, Response and Technology, at the University of New Orleans. Dr. Shirley Laska, founder of CHART, and Dr. Pamela Jenkins, director of the center's Coastal Communities Project, initiated the intellectual process for this project. Dr. Jenkins developed most of the interview guide and believed the concept of "sense of place" would be an important component for coastal residents. Jenkins and Laska conducted informant interviews and soon after put me in charge of most of the interviewing.[3] Over time, it became evident to the team that place attachment would play a primary role in how residents viewed coastal land loss. The question that ultimately became the aim of this book, "What is it like to experience the loss of place?," was born in conferences with Shirley Laska, Pamela Jenkins, and others involved in the Coastal Communities Project. The overall project, however, was more ethnographic in scope, and aimed to capture the daily life and oral histories of community members along with their thoughts about coastal land loss. What I became interested in was their attachment to the coast and how they experienced and understood the loss of their land. Answers arose from residents as we asked them to tell us a kind of story about their life in coastal Louisiana.

One of the answers that residents gave was that they hold a profound attachment to coastal Louisiana. Attachment to places like Louisiana's coast occurs as people gain a felt connection to them. As Pamela Jenkins, Shirley Laska, Traber Davis, and I have noted (2007), this feeling of connection develops through learned personal interactions with place.[4] Gaining perceptions of place while interacting with it in ways that are personally meaningful is shaped by, but not limited to, biography, macro and micro cultural meanings, amount of time spent with the particular location, and the sociohistorical moment. In other words, place attachment is an emotional bond between people and their physical environment based on the gathering of knowledge and experience. In turn, this bond influences behavior (Burley et al. 2007; Altman and Low 1992; Relph 1976; Tuan 1979; Proshansky, Fabian, and Kaminoff 1983). Summing up the experiences that generate attachment, Thomas Gieryn (2000) notes that geographic locales become personally important to us because of "fulfilling, terrifying, traumatic, triumphant, secret events that happen to us there" (p. 481).

Reflecting this attachment, coastal Louisianians were eager to show just how coastal land loss was affecting their sense of who they are. During the time my colleagues and I spent interviewing in these communities, residents made a point of showing us what was happening to them. The efforts by community members to show and tell us about the loss of land are especially significant because we never asked them about coastal land loss; we only told them that we were interested in coastal life. People would take out maps and show us places, parcels of land that no longer exist. Some would show us personal photographs of land lost. Or, as they imparted the loss of place, their homes, they often showed us family photo albums. Some wept. One woman in Lake Catherine, who had written a cookbook which included family and community history as well as documentation of land loss, spent over two hours showing me and fellow researcher Traber Davis a fifty-year-old map of the area where much of the land is now gone. Using his laptop, a council member of the Houma Indian Tribe displayed at least 150 photos of the devastated Isle de Jean Charles community after Hurricane Lili. A seafood dock owner in Lake Catherine took me out in his boat to show me all of the land that has been lost over his lifetime. He was thirty-one years of age at the time. Offers to show us the slow, ongoing disaster were numerous. Many residents were eager. They wanted us to know. They wanted others, outsiders, perhaps those who might be able to get the story to others, to see what was happening. They insisted that "you have to *see* it."

When residents strongly identify with a place and consistent and unrelenting change occurs to place, self-definitions, the components of identity, also change. The ways that residents interpret the disaster of coastal land loss reflect an identity that is contextualized within the unique culture of southern Louisiana. This culture and its identity are changing under the pressure of a mutating landscape. Residents live with coastal land loss and struggle with the meaning of this ongoing change. This struggle was evident during interviews as they revealed an identity that was in question, tenuous. Before Katrina and Rita, people expressed a sense of immediate jeopardy. Even then residents revealed an identity that felt to them increasingly vulnerable and at risk. Without a doubt Hurricanes Katrina and Rita produced further uncertainty about place and the ways Louisiana's coastal residents defined themselves.

However, prior to Katrina and Rita residents viewed the disaster of coastal land loss through lenses shaded by the elements of their identities.[5] In turn, the ways they viewed this mostly slow onset disaster colored the ways they saw place. While what residents said about the places they lived varied, often within individual narratives, land loss always affected conceptions of place. In short, as the disaster of coastal land loss unfolds it mediates how residents view place and themselves (Greider and Garkovich 1994; Brown and Perkins 1992).

As they told their stories, people conveyed an impression of uncertainty about place, the future, and themselves. Their sense of identity was in flux and fragile. It might seem that after Katrina, identity would be wholly dismantled. However, as Kimberly Solet, a reporter for the *Houma Courier*, the primary newspaper of Terrebonne Parish, wrote to me in an e-mail on October 10, 2005, "as you know, because of their strength and determination, many bayou residents are already rebuilding." Certainly, the dose of Katrina and Rita was particularly devastating, but the residents of coastal Louisiana have dealt with similar situations. They have lived with continuous land loss punctuated by storms, some more powerful than others, for generations.

They know how to handle these situations; they know how to rebuild their communities. Their resolve is strong and it has helped them sustain community and their relationship with place over the years. With this in mind, it seems that the determined rebuilding that Ms. Solet mentioned may have served as a coping mechanism. By rebuilding, residents reclaimed a relationship with place and reasserted their identity as residents of coastal Louisiana. Yet there is a new reality of place. It means something different from what it did it did before the storms of 2005. Their relationship to place adjusts as the processes of change continue.

While Katrina and Rita have certainly impacted the identity of millions, Louisiana's coastal land loss remains a personal issue that affects these residents' sense of who they are both individually and collectively. The event is not solely an environmental disaster occurring where they live; it means many different things to these people. An example of this multilayered significance has to do with the fragility that residents experience due to the continual loss of land. As place slowly disappears, residents experience an erosion of security as the places they knew intimately become strange. Identifying as a coastal resident of Louisiana means having a certain familiarity with the landscape. When that landscape changes and becomes strange, that self-definition no longer applies. Further exacerbating community members' sense of fragility is the symbolic loss that occurs with the physical loss of land. Childhood memories, familial connections, and the significance that occupations hold for residents are symbolized by place. As the land disappears, identity is thrown into question due to the erosion of those physical places that played such an important role in forming that identity.

Residents relate to south Louisiana in such a way that it plays a major role in their sense of identity. The voices of the people in this book convey this importance again and again. Because residents hold such a strong attachment and identification with place, damage to their environment becomes injury to the self. In short, metaphorically, residents said that a part of who they are was dying. Residents' sense of identity is in peril and this causes the event to be experienced through anxiousness, desperation, and vulnerability. Put another way, the potential of the event means the possible death of an integral component of identity. Nevertheless, while Hurricanes Katrina and Rita quickly overcame large parcels of land and brought the possible ruination of southern Louisiana closer to a reality, many residents, even those of the most devastated communities, have launched a staunch effort to reestablish place and stave off that potential death of identity. In fact, to imply that residents are the unfortunate victims of outside forces would be misleading to say the least. As their narratives conveyed, they are not only negotiating their identity in relation to coastal land loss. Many coastal Louisianians are acting back against this event and all of its elements in an attempt to shape the process and, as a result, themselves. The rest of this book explores how residents interpret the experience of losing the places where they live.

THE CHAPTERS

Chapter two, "Losing Louisiana: A Meditation on Coastal Land Loss," is the first of two chapters that set the stage for the study. The chapter gives the natural and sociopolitical context for what residents say about the event. I outline the geophysical forces underlying coastal land loss, as best I can for a sociologist, that is. As stated earlier, generally, coastal erosion is a natural process. Waves wash away shoreline, and storms erase large tracts of land instantly. In Louisiana, storms eradicate barrier islands and wetlands, rich ecosystems and protective barriers for coastal communities and cities. However, these natural processes are supposed to be offset by a natural and continual replenishing of land. Human activity in the twentieth century and now into the twenty-first have cut off and strangled this regenerative progression.

After this process is explained, chapter three, "Communal Histories," proceeds to give some historical and political context to this ongoing disaster. Decisions have been made, not made, and behaviors encouraged that have brought the situation in coastal Louisiana to its current point. Furthermore, this discussion is not only important for contextual understanding, but it is the setting where residents' experiences and interpretations take place. The natural, historical, social, and political aspects of land loss have shaped their understandings and negotiations of the event.

Next, "This Is Our Home: Attachment to the Coast," chapter four, launches into the data. In this chapter I present residents' discussion of land loss as it relates to their connections to place. I begin by establishing the theoretical context for residents' place attachment and then outline how places are socially constructed. I discuss how residents, as well as the rest of us, transform physical landscapes into symbolic ones. Community members used their experience of land loss to convey the significance to identity that place held for them. In fact, their attachment to place was the most influential factor shaping their experience and interpretation of coastal loss. As such, I preface residents' voices with a brief outline of how place attachment develops, and the chapter then moves into the utility of narratives in investigating this issue. Next, some methodological information, on the interview guide for example, is summarized in this chapter while more detailed information is given in an appendix. It should be noted that Pamela Jenkins made contributions to theory and method and provided essential guidance in analysis. The remainder, and the bulk of the chapter, establishes residents' attachment to place while placing their narratives within a theoretical context of place attachment and identity.

Chapter five, "Seeing It for Themselves" offers more extensive passages from residents' narratives. Their voices are categorized into themes that expand further on their particular attachment to place. The chapter uses long interview passages to examine residents' understanding of the phenomenon. As in chapter four, *how* residents chose to speak about land loss is a focus. Looking at how community members talk about land loss reveals what it means to them and their identity as coastal residents of Louisiana. As a result, we come to understand what it is like to experience this phenomenon. The themes presented, *the damaging consequences of coastal land loss, restoration, human degradation of the coast, uncertain place, uncertain self,* and *politicized nature*, exemplify how residents characterize land loss. Although themes such as those presented here are constructs of the researcher, they represent narrative commonalities and are useful in explaining the ways people use interpretations to mediate their experiences. Long interview passages are presented to provide as much context as possible, to give room for the reader to ruminate over what people were saying, and so that the point, or meaning, that residents were trying to convey can be better understood. Additionally, it is through community members' extended voices that we gather the essential meaning of their experiences. By ascertaining what residents are essentially saying, we can understand more fully what Louisiana's coastal residents are experiencing and thereby draw correlations with our own, seemingly unrelated, experiences. In short, by moving toward the particular we also move closer to the universal. This chapter is perhaps the most relevant in that it provides a rich description of the varying ways that residents interpret their experience of coastal land loss while expressing the saliency of their attachment to place.

The insights garnered from chapter five allow us to understand how the experiences of coastal Louisianians may inform our own. Equally, these illuminations reveal to us how we might improve upon restoration projects, related policy, and provide some use to coastal community organizations in Louisiana, as well as elsewhere, in anticipation of future needs in other communities. Concluding the book, chapter six, "Saving Place: Residents and Their Environment," addresses not only these issues but the restoration process in general. In this chapter, residents' experience of coastal land loss and the subsequent impact to identity is considered in the context of what that may mean for long-term coastal restoration as well as other environmental restoration processes that we are increasingly faced with. As some places begin to see the direct impacts of global climate change and others its indirect ramifications, the experiences of coastal Louisianians have important lessons for us. In many ways, this may mean "going local." I conclude that it is imperative for residents and their communities to play a vital role in restoration practices, rather than rely on current restoration traditions which all too often cause communities frustration instead of renewal.

2

LOSING LOUISIANA
A Meditation on Coastal Land Loss

In late August of 2005, Hurricane Katrina and then Hurricane Rita, a few weeks later, capitalized on the effects of coastal land loss. More recently, in September of 2008, Hurricanes Gustav and Ike took advantage of the cumulative loss. The storms' impacts were made all the more severe due to more than 1.2 million acres of land lost since the 1930s (LCA 2004). Louisiana's coastal wetlands and barrier islands served not only as a vital ecosystem and resource basin for the nation, but also as a buffer which weakened storms as they moved inland. The land slowly breaks down hurricanes, thereby reducing the impact on more inland communities and urban areas like New Orleans. A healthy Louisiana coast encompasses barrier islands, then hundreds of miles of marshes and coastal wetlands which, perhaps counterintuitively, feel like hard land. Rivers and bayous cut through sections of tall grasses and various tree groves. Human communities dot the landscape all along the waterways. However, Katrina, Rita, and even Gustav and Ike erased large portions of this already weakened land in an instant. Now, Louisiana's coast is more vulnerable to future storms as well as the further slow loss of land.

Prior to Katrina and Rita, when this research took place, issues of environmental change in the region were increasingly coming into the national focus, and coastal restoration projects were being proposed on a scale heretofore unprecedented. Now, in the post-Katrina era, Louisiana's land loss holds a spot on the national stage, yet due to the massive rebuilding efforts that continue to take place along the Gulf Coast, it remains to be seen what actual rebuilding of the coastal wetlands will transpire. What does seem certain is that coastal restoration will receive more attention in the post-Katrina era. Additionally, it appears likely that Louisiana's monumental challenge is indicative of similar environmental issues that particular places will increasingly face as the warming global climate takes localized effects.

Louisiana is faced with an incredible challenge that likely serves as a harbinger for future restoration efforts, because, currently, Louisiana has 30 percent[1] of the nation's wetlands but accrues 90 percent of the country's wetland loss (Louisiana Coastal Area Ecosystem Restoration study [LCA] 2004; Louisiana Department of Natural Resources 2004). Since the turn of the twentieth century Louisiana has lost approximately 30 percent of its wetlands at an average rate of thirty-four square miles per year since the 1950s (LCA 2004; Farber 1996; U.S. Department of Interior 1994).[2] Computer models estimate that an acre of land is lost every fifteen minutes (LCA 2004; American Planning Association 1997). Considering current land loss dynamics and restoration efforts, the loss over the next fifty years is expected to be five hundred square miles (USGS 2005; Louisiana Department of Natural Resources 2004; Barras, Beville, Britsch, Hartley, Hawes, Johnston, Kemp, Kinler, Martucci, Porthouse, Reed, Sapkota, and Suhayda 2003).

Estimates by the U.S. Geological Survey (USGS 2005) revealed that "approximately 100 square miles of wetlands in the Mississippi deltaic plain were transformed into shallow open water by the hurricanes" Katrina and Rita (Working Group for Post-Hurricane Planning for the Louisiana Coast 2006, p. 13). Open water now appears where there once were emergent marsh, areas of unconsolidated shoreline, and floating aquatic vegetation. However, the scientists at the Working Group for Post-Hurricane Planning for the Louisiana Coast state that "it is premature to conclude that these wetland losses are permanent because re-growth from roots and rhizomes and re-vegetation of mudflats may occur during the next growing season or two, as was observed after Hurricane Andrew in 1992" (WGPHPL 2006, p. 13).[3] More recently, it is estimated that these storms wiped out two hundred square miles of the coastal marsh (*WaterMarks* Dec. 2008). While some regrowth has occurred, exact numbers remain unclear, and it is certain that there was a significant loss of land as a result of the storms.

In addition to storms, human action in the environment is a major cause of land loss. A little over a century ago the logging industry in Louisiana boomed as cypress was promoted as a building material. This left a legacy of cypress stumps dotting the backlands (Gramling and Hagelman 2004). As a result of clear-cutting, the wide, deep roots of these trees no longer acted to hold the soil together. At about the same time as the blossoming of the logging industry in the late 1800s, levees were built around the Mississippi River to limit flooding of populated and agricultural areas and to support navigation interests, but not to protect communities from flooding (LCA 2004). This changed after the historic flood of 1927 (Barry 1997). The U.S. Army Corps of Engineers proceeded to levee the entire river south of Baton Rouge (van Heerden 2006). As beneficial to communities and the local economy as this may have been, the levees also eliminated the seasonal flooding that naturally replenished the sediment deposits that built the Mississippi Delta.

The southern region of Louisiana was built over millions of years by an annual spring deposit of sediment that drains into the Mississippi River from its tributaries covering much of the U.S. That sediment would then roll down the river, flood its banks in southeastern Louisiana, and deposit another layer of sediment, eventually compacting and becoming more solid. The following year another layer would be deposited. The levees cut this process off. The land is compacting, but no new sediment is being put in its place. Essentially, the land has been sinking for about a century (Reed and Wilson 2004).

An additional unintended consequence of the levees is that the sediment that is carried down the river each spring now flows out of the mouth of the Mississippi River and spills out into the Gulf of Mexico, bringing with it high concentrations of nitrogen and phosphorous from agricultural activities. This sediment spill-off creates a "dead zone" each summer in the Gulf of Mexico of up to seven thousand square miles from a lack of oxygen in the water (Bruckner 2008). Back in Louisiana's coast, since the leveeing of the late 1920s and 1930s, the natural subsidence that occurs from the loss of sediment deposits has gone unchecked (Reed and Wilson 2004).

Along with leveeing the Mississippi River and cypress logging, another resource extraction activity contributed to the state's land loss. The oil and gas industry was welcomed into Louisiana as a sort of economic savior beginning in the 1930s. There were consequences, at best unintended and at worst ignored, resulting from mobile drilling rigs that left wide channels from canals that were dug for access, exploration, transportation, and the laying of pipelines. An estimated ninety-three hundred miles of pipeline zigzag across Louisiana's coastal wetlands (LCA 2004). These canals and channels allow salt water from the Gulf of Mexico into freshwater marshes, thus destroying vegetation (LCA 2004). Saltwater intrusion created by these canals along with breakwaters and segmented barriers have also undermined the vast network of barrier islands that encapsulate and help regulate a unique mixture of freshwater and brackish wetlands. Barrier islands also serve as a first line of defense against storms (van Heerden 2006). According to coastal geologist Ivor van Heerden (2006), these islands will disappear totally if coastal land loss continues at its current rate.

Researchers generally agree that the canals cut by the oil and gas industry are a major cause of wetland depletion and account for as much as 69 percent of all wetland loss (Reed and Wilson 2004; Hecht 1990). In addition, runoff and pollution from exploration and extraction make the problem worse by killing vegetation, inducing chemical transformations, and altering sediment transport and the migration of organisms (LCA 2004; Hecht 1990).

Canals also create "spoil banks"—the dredged material placed adjacent to the canal. These banks create land much higher than the natural marsh surface and alter the flow of water across wetlands. Coastal geologists Denise Reed and Lee Wilson (2004) state that "canal dredging [most of which occurred between 1950 and 1980 (LCA 2004)] altered salinity gradients and patterns of water and sediment flow through marshes and not only directly changed land to open water, and marsh to upland, but also indirectly changed processes essential to a healthy coastal ecosystem."

Reed and Wilson (2004) argue further that natural forces of land loss did not change during the mid- and late twentieth century but that the accelerated loss during this time was due to human action upon the environment. Scientists concur that a major cause of all land loss is human induced (Barras et al. 2003; Penland, Wayne, Britsch, and Williams 2002). While natural factors, which include wave action, storm surge, eustatic sea-level rise, and geological compaction, have not significantly changed during the past century (Reed and Wilson 2004), the effects of these are multiplied when there is no land compensation and saltwater intrusion is expedited (van Heerden 2006).

Most of Louisiana's land loss occurs inland as wetlands turn into open water. This differs from typical coastal loss that occurs at the shore such as the erosion that coastal California experiences (Penland et al. 2002; Hecht

1990). Furthermore, current forms of oil and gas mining are being linked to the continual regionwide loss. Robert Morton's and colleagues' (2002; 2003) continuing research for the USGS finds more and more correlation between the extraction of oil and gas and land subsidence. While it was commonly thought that the oil and gas under Louisiana was so far underground—eleven thousand to eighteen thousand feet—that it would not effect subsidence, Morton's hypothesis of "regional depressurization" caused by oil and natural gas extraction has gained wider acceptance (Morton, Tiling, and Ferina 2003; Morton, Buster, and Krohn 2002; van Heerden 2006).

While the oil and gas companies dispute the claims by Morton and his colleagues, they, along with all of the other industries with investments in the region, acknowledge the magnitude of the problem of land loss. This is a fairly recent development, and, although seemingly not aware of the full ecological importance of this ecosystem, private industry and government are sensitive to the region's economic value. One hundred million tons of cargo is shipped annually through the waterways. The region is home to a fishing industry that contributes $2.8 billion a year to the state and national economies. Thirty-four percent of the country's natural gas and 29 percent of the nation's crude oil supply comes through or from south Louisiana (LCA 2004). The region further serves as the biggest domestic source of oil, pumping out more than even the Alaskan pipeline. Due to subsidence, however, much of the pipelines that used to be buried in marsh and wetlands now lie exposed above ground (van Heerden 2006).

Additionally, the beach at Port Fouchon, a major staging area of the oil and gas industry, loses forty feet a year. The average beach erodes only a few feet each year, according to van Heerden (2006). Louisiana Highway 1, the only current road that connects Port Fouchon to the interstate highway system and New Orleans sixty miles to the north, floods frequently from mild storms or even sustained southerly winds (van Heerden 2006). Highway 1 is also the only road that goes in and out of Grand Isle, one of the communities highlighted in this book. This sort of vulnerability has garnered the attention of private industry and government who now recognize the economic importance of the region's ecosystem. Consequently, conservation and restoration plans have been widely proposed over the last decade (Reed and Wilson 2004; LCA 2004).

For urban centers like New Orleans, the coastal marshes and wetlands once offered protection from tropical storms and hurricanes (Bartell et al.

2004). A storm tide pushed inland by hurricanes falls a foot for every three miles of marsh it must cross (LCA 2004). But as the land disappears these storms retain much of their strength, moving farther inland and inundating interior wetlands with salt water from the Gulf of Mexico (LCA 2004). As for New Orleans, the city is seventeen feet below sea level in some places (LCA 2004). Hurricanes Katrina and then Rita exploited the city's vulnerability—its position below sea level and its unprotected status due to cumulative land loss. Katrina and Rita have further exacerbated land loss by overcoming already weakened wetlands with salt water from the Gulf of Mexico.

Indeed, NOAA (2005), USGS (2005), and the Working Group for Post-Hurricane Planning for the Louisiana Coast (2006) originally reported that the storms eradicated nearly one hundred square miles of coast, with more recent estimates reaching two hundred square miles of loss (*WaterMarks* 2008). NOAA in particular conducted an "immediate post-storm assessment" of land loss using their Coastal Change Analysis Program (C-CAP) (NOAA Coastal Services Center 2005). Comparing poststorm data to information collected in 2001 as part of C-CAP, NOAA acknowledged that some of the land loss simply occurred over the four years, but most was due to Hurricane Katrina (NOAA Coastal Services Center 2005). What this means for New Orleans is that a category three storm, as was Katrina at landfall, has a greater impact than a similar storm in the past because of land that has disappeared. In other words, the strength of a storm may be the same across time, but the increasing loss of coastal land means the impact will be greater.

However, prior to Katrina and Rita, coastal land loss had actually been on the decline in recent years. The deceleration of land loss might seem like a bit of good news, but this may be deceptive. David Chambers of the Louisiana Division of Environmental Quality says that decreasing loss is because the most vulnerable land is "gone, and there's nothing left to lose" (Hecht 1990, p. 40; also corroborated in Reed and Wilson 2004 and Morton et al. 2002). It seems that a theory of diminishing returns developed by Woody Gagliano and Robert Morton and his colleagues (2002) is in play here. They argue that fault blocks (large portions of the earth's crust) are contributing to subsidence. These faulting events take place when existing geological faults are activated by lower underground pressure caused by the withdrawal of oil and gas. Much of the inland oil and gas has already been removed. Thus, fewer faulting events are occurring. However, the U. S. Army Corps of Engineers report in the LCA (2004) that inland areas will continue to succumb to saltwater intrusion through a combination of storm surges, continued resource extractive activities, and subsidence.

As if all of this weren't enough, global climate change is compounding Louisiana's land loss. In fact, two years before Katrina and Rita coastal scientists were saying that Louisiana was "more vulnerable to the effects of climate change than any other part of the U.S." (*WaterMarks* Feb. 2003, p. 5). Biologist Robert Twilley says that the coast is sinking at the same time that the sea level is rising. This is happening now, not in some distant scenario. Twilley refers to a "two inch relative sea-level rise in some regions over the last ten years," and over the next century it could range from fifteen to forty-four inches (*WaterMarks* Feb. 2003, p. 20). Storms will have greater impacts if not from higher intensity then certainly from sea-level rise. The acceleration of land loss due to climate change will also threaten communities, navigation, industry, agriculture, and fishing.

In light of all of this, the post-Katrina era presents a historic moment for acting. As Twilley optimistically and realistically puts it, the good and bad news is that we are part of the problem (*WaterMarks* Feb. 2003). To begin with, implementing projects can stave off further loss and replenish elements of the ecosystem (Morton et al. 2002, 2003; Reed and Wilson 2004). In rebuilding the wetlands, one of the top priorities is to reconnect the river system to the marshes (Reed and Wilson 2004). The state is looking for substantial assistance and has developed a large-scale plan to make their case at the federal level (LCA 2004).

In 2004, the LCA or Louisiana Coastal Area Ecosystem Restoration plan (LCA 2004) was submitted to the federal government by the U.S. Army Corps of Engineers. The plan called for an ecosystem-wide restoration project costing $14 billion. It was deemed way too expensive and didn't make it past the Office of Management and Budget. In a separate piece of coastal restoration legislation the Bush administration recommended $1.9 billion to focus on implementing five projects and studying ten more over the next ten years (Schleifstein 7/31/05). Relief came to LCA proponents in the form of the 2006 U.S. Energy Appropriations Bill, which allows Louisiana to collect revenue from oil extraction occurring off its coast, revenue that the state was not allowed in the past (Alpert 11/30/07). Louisiana, prior to Katrina and Rita, was to collect $540 million for restoration projects (Schleifstein 7/31/05). Now, due to the Water Resources Bill passed in late 2007, the state can move ahead with its plans to use another $255 million in federal money for more conservation projects (Alpert 11/30/07). The major projects that this collective funding would support are acknowledged by all involved as only a starting point, but LCA projects are getting closer to implementation even while meeting with continued resistance.[4] Projects funded by the above monies include work along the Mississippi River Gulf Outlet, efforts for shoring up coastal Jefferson, Lafourche, and St. Bernard parishes, and the studying of a larger plan to redesign the mouth of the Mississippi River (LCA 2004).

Much of this funding would go towards conventional coastal restoration projects, which generally take three forms—shoreline protection, hydrologic restoration, and wetlands creation. Of these, wetlands creation is considered to be the most productive, sustainable, and least detrimental (van Heerden 2006). However, recently many have been advocating hydrologic restoration through sediment delivery systems. These delivery systems entail dredging sediment from riverbeds and transferring that sediment, called "slurry," through large pipelines to strategic areas targeted for wetland rebuilding.

Nonetheless, as it stands currently, wetland re-creation occurs primarily through river diversion projects. Essentially, holes are cut in the levee of the Mississippi River, for example. Locks are then put in place. Then, as the name implies, sediment-rich water is diverted from the river into specific parts of the wetlands. The locks serve to stop, start, and regulate the amount of water diverted. In addition to sediment deposition, the diverted fresh river water also reduces saltwater intrusion and helps to reach favorable salinity of the wetlands and waterways allowing for further wetland creation.

Two major river diversion projects are currently in operation—Davis Pond Diversion on the west bank of St. Charles Parish and the Caernarvon Freshwater Diversion on the border of St. Bernard and Plaquemines parishes. These two projects were largely funded by the 1989 Coastal Wetlands Planning, Protection and Restoration Act (CWPPRA), which is also known as the Breaux Act after Senator John Breaux.[5] It has received about $40 million a year. The legislation has limited credibility, however. Some of the neediest areas and projects don't get funded, it receives stiff competition from developers and land owners, and what seems like a dubious project evaluation process to scientists like van Heerden greatly diminishes the effectiveness of CWPPRA. Additionally, as almost all of Louisiana's wetlands are privately owned, some plans develop problems as landowners object to restoration projects that run through or on their property. Notwithstanding these obstacles, many projects have come to fruition, although most are small-scale.

Even Davis Pond and Caernarvon are meant to display the potential of such restoration projects and provide justification for future funding. Up to this point however, they have only operated at 10 percent of their capacity (van Heerden 2006) due, in part, to struggles over fishing rights and development.

Adding to the limited effectiveness of CWPPRA, coastal scientists and environmentalists say the LCA projects to be funded by offshore oil and gas royalties are jeopardized by different land use decisions (Schleifstein 7/18/04). Before Katrina and Rita, activities that undermined coastal restoration were allowed to persist largely unabated in southeast Louisiana. State and federal agencies continued to approve development for business, housing subdivisions, and recreational camps, as well as other activities such as logging that intensified land loss in the very areas that the projects were designed to restore.

For instance, continued suburban development in the region has required draining the wetlands in order to build. This removal of water from the soil depletes its dense substance and also increases the oxygen content, further adding to its decomposition (van Heerden 2006). John Day, chairman of the National Technical Review Committee, a group of scientists providing an independent review of the LCA program, said that the proposal would not provide any significant restoration (Schleifstein 7/18/04). According to Day, the projects that have the greatest chance of reversing land loss were left out of the proposal. He added that "if Louisiana is serious about putting large resources into coastal restoration, it can't go around destroying the very resources it says it wants to preserve" (Schleifstein 7/18/04). Col. Peter Rowan, chief of engineers in the U.S. Army Corps of Engineers' New Orleans regional office, provided the counterpoint, saying that "this has got to be a balanced process. Why restore the coast if people can't live and work in it?" (Schleifstein 7/18/04). These differences in what coastal restoration is and how it should proceed highlight an ambiguous future for the restoration of Louisiana's coast.

The post-Katrina era brings new awareness and consideration; however, questions about how much will change appear as uncertain as ever. Like most issues in our society, Louisiana's coastal land loss is a problem that is not easily rectified because it is multifaceted and many people see its causes and solutions quite differently. Although land loss in Louisiana is, in large part, tied to economic needs and political decisions, these choices take on varying shades according to the perspectives of scientists, engineers, environmental activists, politicians, business leaders, and those favored by the media. What is left out of this milieu are the views of the region's residents. Residents' voices are often overpowered and get lost under the weight of the economic, political, and scientific discourse. And, of course, their voices are part of this larger mix, shaping and shaped by an ongoing history.

COASTAL LAND LOSS AND THE PLACE OF LOUISIANA'S OIL AND GAS ECONOMY

As communities come to decisions about how to use the coast, certain ambiguities arise. The oil and gas industry is a large part of Louisiana's economy, and it is also an activity immersed in the coastal wetlands, making it a part of how residents experience the phenomenon of land loss. Given its great contribution to land loss and its history in Louisiana, which dates back almost a century, the industry serves as a key avenue through which residents experience their world.

Freudenberg and Gramling's (1994) work on the dichotomous perceptions between Louisianians and Californians on offshore oil and gas drilling provides the historical, biophysical, and social backdrop for which drilling was welcomed in Louisiana. Their analysis is useful here not only because oil and gas exploration plays a significant role in Louisiana's land loss, but also because, from an economic standpoint, oil and gas exploration has shaped the symbolic meaning residents have attached to place both prior to and now in the midst of extensive coastal land loss.

Louisiana's oil development arose during the 1930s and 1940s, a period that was marked by an "unprecedented faith in technology nationwide" (Freudenberg and Gramling 1994, p. 75). Realization of environmental destruction due to unabated technological expansion was still decades away. Industrialization and its promises of a better life were welcomed in many areas of the country, but especially in an area such as coastal Louisiana, where the only other major industry was the harsh lifestyle of the fisheries. Along with its gradual growth, the industry further established itself in the region by involving locals, which gave residents a sense of pride during an age when industrialization and technological advancement were signs of great progress.

In addition to the oil industry's economic influence, residents' perceptions of place had been shaped by the biophysical environment. Most of

Louisiana's coastal residents live somewhat inland; that is, they don't live directly on the coast, as is likely in many other coastal areas in the U.S. Most coastal residents of the country can drive along the coast and view it directly, as in California where residents and visitors alike drive along its famous Pacific Coast Highway. As a result, they view the coast as "both a resource and important recreational feature" (Freudenberg and Gramling 1994). On the other hand, most people don't "see" the actual coast of Louisiana, because the marshland hides it from view. There are few beaches, and the ecosystem can be harsh to humans with "more mosquitoes and alligators than spectacular visual imagery" (Freudenberg and Gramling 1994, p. 79). All in all, Louisiana's coastline, prior to the onset of coastal land loss, was low in social salience due, in part, to its inaccessibility to most of the land-based population (Freudenberg and Gramling 1994, p. 88).

Along with the historical and topographical components noted above, social elements in Louisiana's population fostered the fairly easy acceptance of oil and gas activity. First, Louisiana's consistently low educational levels, especially low in rural coastal areas during the 1940s, made entry easier for a large new industry promising to raise the standard of living. Also, there was little conflict or competition between the oil industry and the fishing trades, the other major extractive activity in the region. By contrast, when examining why Californians rejected and Louisianians accepted the oil and gas industry, Freudenberg and Gramling (1994) noted that those residents involved in an extractive industry such as fishing, as opposed to those involved in manufacturing or service industries, are less likely to object to a new extractive industry such as oil and gas, which does not seem to compete with the traditional activity.

The likelihood of opposition in coastal Louisiana was further reduced because of the abundance of marshland. The marsh provided adequate harbor space for both industries, as well as the unintended consequence that the oil rigs came to serve as artificial reefs around which fish gathered. Louisiana's coastal waters lack natural reefs (Freudenberg and Gramling 1994). A third social factor encouraging the acceptance of the oil and gas industry into Louisiana was interaction patterns. While many along the coast, and across the state for that matter, did work in the industry, people who didn't were very likely to have friends and family who did, and still do, and this could be expected to affect their attitudes (Freudenberg and Gramling 1994).

Not only did the slow influx of the industry provide jobs, but it brought an incredible economic boom to the coastal economy during the 1970s and early 1980s (Freudenberg and Gramling 1994). This, coupled with the relatively small number of environmental accidents, served to foster the perception that oil and gas presented low levels of risk.

The historical, biophysical, and social components noted above not only explain the ease of acceptance of the oil and gas industry into coastal Louisiana but also explain, in part, how those residents conceive of place. The favorable reception of oil and gas was influenced by the above factors that were shaped by the reciprocal social construction of these characteristics. These factors were a sort of feedback loop of mutually reinforcing narratives between the oil industry and residents where the industry did not appear to present a risk and, in fact, brought visible gains to a place that was in need and held a ready-made acceptance of industry. The oil and gas industry, while not static, is a fairly fixed component of place, and its evolution in Louisiana helps to explain how residents welcomed an industry that they would later learn produced a large negative environmental impact. We can even see how the incoming industry helped to promote further attachment to place, bringing economic gain and development to an already culturally unique place.

Of course, the above only tells a part of the story of oil and gas in southern Louisiana. And for the purposes of this book this is all that is necessary. Nonetheless, there is the social history that not only fostered the development of oil and gas as a primary cylinder of the state's economic engine, but also shaped and was shaped by its residents.

3

COMMUNAL HISTORIES

The communities within the parishes from which interviews were collected—Jefferson, St. Bernard, Terrebonne, Plaquemines, Orleans, and Iberia parishes—have long histories, with some existing more than two centuries since European settlement.[1] Over this time, these communities have always faced some sort of change due to hurricanes, development, and erosion. However, the unabated disappearance of coastal land over the past two generations was part of the overall physical and social context in which residents developed their conceptions of place. Their attachment to coastal Louisiana and their specific communities developed in a particular historical context, which in the case of this study was captured before Hurricanes Katrina and Rita. That was a historical moment much different from what exists now. Nevertheless, the historical moment present in 2002 and 2003 necessarily developed from the past just as the present moment stems from Katrina, Rita, the sociopolitical aspects of those storms, and the historical moments that preceded them.[2]

GRAND ISLE

Existing within Jefferson Parish, Grand Isle was the first community studied (see the map, figure 1, of coastal communities in the photo insert). Twenty interviews were gathered in Grand Isle, an island that sits off the southeastern coast and is the only occupied barrier island in Louisiana. The island community also has a unique history and has been a popular resort destination since the 1850s.

The island was settled in the eighteenth century with small Spanish outposts. It became famous for its beginnings as a buccaneering community headed by Jean Lafitte. However, after the Louisiana Purchase and an American takeover in 1803, the U.S. Navy terminated Louisiana's coastal privateering in 1814, thus ending this period in the island's history (Reeves 1985). Grand Isle then shifted to sugar production and a plantation economy. Following the Civil War and the elimination of slave labor, the economy moved more toward resort development made possible by a growing national trend in leisure travel and the construction of a railroad to the island. With the ease of regular transportation, Grand Isle became more accessible to New Orleanian elites who looked to escape the summer heat and urban disease outbreaks like yellow fever (Steilow 1981; Reeves 1985).

This era, the mid-to-late 1880s, is often referred to as the Golden Age of Grand Isle, and it is this period that authors Kate Chopin and Lafcadio Hearn celebrate in American literature. This Golden Age came to an abrupt end in 1893 due to one of the deadliest hurricanes in Louisiana history, a storm which destroyed most of the island's structures and which occurred before storms were bestowed with names (Meyer-Arendt 1985; Reeves 1985; Steilow 1981).

After the storm, the island's permanent residents reestablished their livelihoods through the traditional methods of fishing and farming. From then until World War II, tourist development on the island was piecemeal. Following World War II, however, tourism once again began to grow steadily, accompanied by a period of industrialization due mostly to oil exploration, a boom that, as discussed in the previous chapter, lasted until the early 1980s. A bridge from the mainland designed for automobiles was completed in 1934, and post–World War II tourist expansion centered around the development of summer homes along the shore, which contrasted with the islanders' preference for more protected wooded areas located toward the center of the island (Meyer-Arendt 1985). Along with the more traditional occupations in the fisheries, locals and the newly arrived took advantage of the wave of industrialization which was concentrated in the oil business, and by 1962, 134 oil wells were operating on the island block (Steilow 1977).

But the boom was not to last. In the 1980s, the island suffered an economic decline due to an oil bust (Gramling and Freudenberg 1990). Following this economic collapse, localities throughout south Louisiana began emphasizing tourism by promoting the uniqueness of the local Cajun culture (Ancelet 1989).

Thanks to this shift in emphasis, the 1990s brought a resurgence of tourist activity to Grand Isle. New growth had been stimulated by the completion of a waterline to the island, and yet, even though Exxon retained a refinery and business operations on the island, few locals were employed there and all drilling was moved further offshore. In addition, over the last few decades commercial fishing has been in steady decline, leading the local economy to rely largely on tourism (Thompson 7/31/2002).

As of the 2000 census, Grand Isle had 1,541 residents and, like all the communities under study, was defined as rural (U.S. Census 2000). The dominant ethnic heritage is Acadian, but also includes French, Portuguese, Spanish, Italian, and Filipino settlers from its buccaneering and plantation days.

Along with its unique cultural features, Grand Isle holds a geographical quality that is distinct from the other communities. Grand Isle, a barrier island nearly eight miles in length and a mile wide, is on the front line of storms. On the other hand, unlike many of the other communities and precisely because it is an island, the town is above sea level. Thus, while it experiences significant damage during storms, water recedes quickly. Such appears to be the case with Hurricane Katrina. The island suffered extensive wind damage but was not totally devastated and the water withdrew quickly. Because of its southeastern location in Louisiana, Grand Isle saw minimal effects from Rita, which made landfall in southwestern Louisiana and Texas.

SOUTH TERREBONNE PARISH

While Grand Isle was chosen because of its unique history and geography, Cocodrie, Dulac, and Chauvin, in southern Terrebonne Parish, are a small cluster of communities that share a geographical and socially interdependent space. Twenty-three interviews were gathered in these southern Terrebonne communities.[3]

Immigrants from Asia, Africa, the Caribbean, and Spain shaped this area's history. But beginning in the nineteenth century, two groups became dominant: exiled French Acadians from Nova Scotia who came to be known as Cajuns, and Native Americans known as Houma Indians who had been pushed southwest from Mississippi and Alabama by advancing U.S. settlement. Yet today, Chauvin and Cocodrie, with a combined population of 1,723, are self-identified as white (U.S. Census 2000). Dulac's population hovers around 2,000 and is almost 60 percent Native American, the other 40 percent being white (U.S. Census 2000).

South Terrebonne's environment is not as diverse as its population. With more than 90 percent of Terrebonne Parish being wetlands, residents live within a true wetlands environment. As a result, "traditional vocations in Terrebonne Parish are fishing, hunting, trapping, sugarcane farming, boat building," and since the 1920s, oil and gas production (History and Geneology in Terrebonne Parish, LA 1997). But during the 1980s Cocodrie, Dulac, and Chauvin, like the rest of the state, suffered from the oil and gas bust. While they do not currently hold any wells or refineries, the communities still have a substantial number of residents employed by the industry.

These communities also endured a major setback from Hurricane Andrew in 1992. While recovery had been substantial, structural effects of the storm lingered, in part due to the decline of the fishing industry where commercial fishers watched a steady drop in price for their crop over the last twenty-five years. Despite these setbacks the parish accounts for 25 percent of the state's seafood production, ranks first in Louisiana in natural gas production and third in oil, and was able to rebuild its shipbuilding industry upon the demand for gambling boats (History and Geneology in Terrebonne Parish, LA 1997).

Gambling boats bring tourism, and while that industry has never been a part of Dulac or Chauvin, over the past twenty years or so tourism has become an increasing element of Cocodrie's development. Another element unique to Cocodrie is a large marine and coastal research, education, and public service institution—the Louisiana Universities Marine Consortium (LUMCON). The organization, formed in 1979, joins thirteen state universities and higher education management boards. The institution has become a magnet of sorts for local communities providing assistance to fishermen and holding year-round educational programs for schoolchildren from across the state.

LUMCON and the communities of south Terrebonne felt the western brunt of Hurricane Katrina and the far eastern edges of Hurricane Rita. They experienced significant flooding and wind damage, but not widespread devastation. And although much of its housing stock was damaged and federal help was slow to come, residents returned quickly and began repairing as they could. The residents of Terrebonne were spared again during Hurricane Gustav. Two days after the storm, the parish sheriff, Vernon Bourgeois, reported that the majority of housing was intact and suffered only minimal damage (NPR, Sept. 3, 2008). They were not so lucky with Ike, which struck some days later and seemed like a replication of Rita.

EASTERN ST. BERNARD PARISH

Unlike in south Terrebonne, many of the communities of eastern St. Bernard—Delacroix, Yscloskey, St. Bernard, and Toca—were obliterated by Hurricane Katrina. Katrina dealt St. Bernard Parish a devastating storm surge which destroyed key levees. The eastern portion of St. Bernard is the most rural and coastal, making it vulnerable to storms. Its geographical position is also one of the reasons it was chosen for study by my superiors, and thirty interviews were conducted there. St. Bernard is adjacent to New Orleans and extends in an eastern direction to the Gulf of Mexico. An accurate land map shows that the bulbous portion of the parish is all wetlands, and communities begin just inland of this. Archeological findings of Native American complexes along the parish's ridgelines trace the parish's social history back two thousand years. However, for reasons unknown, they had abandoned the delta prior to the arrival of the Europeans (LSU AgCenter 1998).

On February 2, 1699, Iberville, Bienville, and the first French settlers landed on the Chandeleur Islands on the eastern shore of the parish. However, they did not remain, and it was not until 1778 and the Spanish Colonial period that St. Bernard began to be settled by Europeans. Many land grants were given to Spanish families who identified themselves as Isleños, or islanders who immigrated from the Spanish Canary Islands. Numerous descendents of these families still reside in St. Bernard (LSU AgCenter1998).

The Spanish settlers rooted the parish in agriculture, which gave great support to New Orleans. Fishing, fur trapping, and farming, especially of sugarcane, in which Italian immigrants specialized, were the major modes of resource extraction. To this day much of St. Bernard is populated by Spanish, French, and Italians (LSU AgCenter 1998).

St. Bernard is also the locale of significant American military history. In 1814 Andrew Jackson, with a ragtag band of regulars, volunteers, and pirates, defeated British troops who had invaded from Lake Borgne and the eastern end of the parish. And on April 12, 1861, P. G. T. Beauregard, a native of St. Bernard, gave the order to bombard Ft. Sumter, plunging the North and South into civil war (St. Bernard Parish Library; LSU AgCenter 1998).

Despite the parish's active military role, it remained agriculturally based until after World War II, when oil and natural gas became dominant. As a result, while commercial and sport fishing are still major elements in St. Bernard, farming is greatly diminished. By 1970 oil and gas production had become the primary employer in the parish, and the growth of the industry, along with the parish's proximity to New Orleans, facilitated great suburban expansion which, until Katrina, continued to move eastward toward the more rural areas of the parish (LSU AgCenter 1998).

In 1965 the state finished construction of a man-made canal that cut through the entirety of the parish to serve as a shortcut for shipping from New Orleans to the Gulf of Mexico. The canal, the Mississippi River Gulf Outlet (or as commonly referred to, MRGO, pronounced "Mr. Go") all but put an end to the parish's fur production, as it destroyed freshwater marshes by allowing saltwater intrusion from the gulf (LSU AgCenter 1998).

MRGO has been a source of controversy since its inception. The conflict between the Port of New Orleans Dock Board and St. Bernard officials began as the decision to build MRGO was reached by the dock board, a decision in which St. Bernard felt they had little representation. St. Bernard believed the canal was forced on them, and officials have been trying to have the channel closed since its opening. Until 2007 the Port of New Orleans Dock Board and the U.S. Army Corps of Engineers, who make primary decisions about the canal, continued to argue for MRGO as a way to ease shipping even though the canal was often not open at full capacity due to the frequent need for dredging. The parish government and residents constantly fought the U.S. Army Corps of Engineers and the Port of New Orleans Dock Board, claiming that the canal brought no economic progress as promised and only caused frequent residential flooding and massive coastal land loss (LSU AgCenter 1998).

Residents and parish officials alike feared their communities were more vulnerable to land loss and storms due to MRGO. Their fears came to fruition on August 29, 2005, as the communities of eastern St. Bernard fell under the eyewall of Hurricane Katrina. The communities included here were almost completely annihilated. At this point, although the future remains uncertain, rebuilding is slowly occurring. And finally, in late 2007 the U.S. Army Corps of Engineers has agreed and even expedited plans to close the notorious MRGO.

LOWER PLAQUEMINES

Plaquemines, the southernmost parish in Louisiana, is a ninety-mile-long peninsula stretching out from the southeastern portion of the state as the Mississippi River spills into the Gulf of Mexico. Small communities revolving around resource extraction sit along two strips of high ground bordering the river of this delta region. Citrus, sulphur, oil, commercial fishing, and fur trapping have been the primary industries of the parish. In 1946 orange growers and parish officials organized the first Orange Festival, which, despite decimation from freezes and hurricanes, has nonetheless grown into the popular Plaquemines Parish Fair and Orange Festival held annually in December (Louisiana Collection—UNO Library).

The Plaquemines citrus festival is but one example of the social history that evolved out of its resources. The parish, like that of St. Bernard, dates back more than two thousand years, and the earliest known culture is the Tchefuncte. Europeans first traversed the region in 1682 as René-Robert Cavelier de La Salle proclaimed the whole of the Mississippi Valley as property of the king and queen of France. Eighteen years later, in 1700, Iberville built the first French fortification along the Mississippi River in Plaquemines (Louisiana Collection—UNO Library). But, as in the rest of southeastern Louisiana, the French were not the only migrants to gain a foothold in the region. The parish's cultural history comprises French, Spanish, Slovenian, Dalmatian, Chinese, Filipino, African American, and, since the 1970s, Southeast Asians.

While Plaquemines may have diverse cultural influences, its political history is more singular. From 1920 to 1980 the parish's political structure was dominated by the Perez family and their now-notorious patriarch, Leander Perez. Although his official position for most of his career was district attorney, he wielded great power. Perez considered lawmakers "superfluous middlemen. He simply drew up laws and inserted them into the minutes of the parish police jury and commission council. He concluded that honest elections were more trouble than they were worth and made sure none was made in his bailiwick" (Jeansonne 1995, pp. xiii, xiv). The Plaquemines marshes were found to be oil rich in 1933, and Perez set up corporations to lease the public land from the parish, which then leased the land to oil companies. Perez was paid by both the parish as a district attorney and the oil companies as landlord (Jeansonne 1995). He became an oil tycoon worth millions who wielded unparalleled power in Plaquemines, power used for his own advancement while socially, economically, and environmentally pillaging the parish. In 1980 the parish redistricted and in 1982 the state attorney general superseded the local district attorney's handling of oil leasing (historically a Perez), thus bringing the sixty-year Perez political dynasty to an end (Jeansonne 1995).

With the decline of the Perez family's power, the communities of Buras and Empire received an influx of Southeast Asian immigrants who were drawn to the fishing industry in the late 1970s and 1980s. Slovenians and Dalmatians, who traditionally engaged in the large oyster harvesting of the parish, were now joined by Southeast Asians, which resulted in increased competition and some ethnic resentment.

In the years leading up to Katrina, the oyster industry had become the source of much controversy. Oyster harvesters leasing land from the state brought lawsuits against the state for damages to their oyster beds by siphons and river diversions used for coastal restoration. While some claims were legitimate, many appeared questionable, because some of the land under dispute had never been cultivated for oysters, much less harvested. Many plaintiffs were also accused of leasing oyster beds upon hearing of the impending lawsuit. A large decision in favor of the harvesters by a St. Bernard judge threatened to cost the state $700 million and lead to the view that the suits were undermining the coastal restoration process (Meitrodt 5/04/2003). The ruling by the St. Bernard judge was appealed by the state, and in May of 2005 the U.S. Supreme Court rejected the awards to the oystermen, essentially ending an eleven-year dispute (*Times-Picayune* 5/24/2005).

But the courts couldn't stop Hurricane Katrina, and although the storm hurt the 2005 oyster harvest significantly, the yield produced more than what many had projected. Even so, the communities of lower Plaquemines, like those of St. Bernard, were completely wiped out by the storm. The parish's southern communities, where twenty interviews were gathered, are rebuilding slowly. Progress is piecemeal, and there are plans to consolidate community services into one municipality that will serve the southern region of the parish.

DELCAMBRE

Such controversy around coastal land loss or restoration does not exist for the community of Delcambre, where fifteen interviews were collected. It is only since Katrina and Rita that they have felt the need to become more involved in these issues. This community of about two thousand sits on the border of two southwestern/central Louisiana parishes—Iberia on the east and Vermillion on the west. Delcambre was initially chosen for study as a comparison community and for its similar demographics to the eastern communities.[4]

The community developed, as the region did, primarily through the influx of Acadians who were expelled from Nova Scotia by the British. The French also settled in the area, first arriving in the early 1700s. While the region shifted to Spanish rule in 1762, it retained a strong French influence (Iberia Parish Tourist Commission).

These settlers established the community's traditional industries including fishing, hunting, trapping, and farming. Sugarcane farming came to dominate the area and still has a strong foothold. The natural gas and oil industry is also a major business here but like the rest of the state has seen a substantial decline in investment. Facilitating that decline was a disaster in 1980 at the Jefferson Island salt dome. Nearby Lake Peigneur was drained by a gas rig puncture of the salt mine, nearly killing several workers. Since, the parishes have sought court injunctions to keep oil and gas companies from resuming drilling near the salt mine (Schultz and Simoneaux 8/25/1994).

Commercial fishing in the community has also greatly declined. The area comprises less wetlands than the other communities, and farming has always been more prominent than commercial fishing. Likewise, considerably fewer coastal restoration projects have been implemented here than in the southeastern region of the state where land loss occurs at its highest concentration (Schultz 10/25/1996).

Interestingly, despite Delcambre sitting on Louisiana's central coast west of Hurricane Katrina's landfall, it still experienced significant flooding, which was compounded by a fierce blow by Hurricane Rita. The extent of the damage was widespread and many remain displaced, but rebuilding is gaining ground.

LAKE CATHERINE

Lake Catherine suffered a similar fate from the storms of 2005. This community, where eighteen interviews were conducted, lies on the eastern shore of Orleans Parish. Lake Catherine is surrounded by Lake Borgne, Lake St. Catherine, and Lake Pontchartrain. Unlike the other southeastern Louisiana communities in this study, it is not directly connected to the Gulf of Mexico. Lake Catherine is in the midst of the Bayou Savage National Wildlife Refuge thirty minutes from the New Orleans city limits.

Like the refuge, Fort Pike is another treasure of this community. Fort Pike, completed in 1826, was joined by nearby Fort McComb as part of a large-scale coastal defense system along the gulf and Atlantic coasts. These forts were designed to help protect ports from land or sea invasion (Louisiana State Parks 2008), and while Fort McComb is now gone, Fort Pike remains as a state historic site and a popular attraction for those visiting New Orleans.

While the forts represent a historic public interest, Lake Catherine has largely been under private control for most of its European and American history. First granted to Gilbert Antoine de St. Maxent in 1763, Lake Catherine remained under single ownership for the next two hundred years. During that time the land came to be owned by several prominent New Orleanians. Throughout, a community of small homes and fishing camps was built on land leased from the owner.

CSX Railroad was the owner in the 1980s but in 1989 decided to sell the land rather than install a sewerage system required by the state. The railroad sold the land, divided into four parcels, to the residents for about $2.5 million, and they organized as the Lake Catherine Land Corporation. The company leased the land to its shareholders—the residents. The company was formed because residents were not able to gain title to their land until a state-approved sewerage system was in place for the community. The state demanded a full-scale treatment plant but did not want to pay for a large-scale system itself. The residents argued that if the state would not foot the bill, then they should be allowed to install individual treatment plants. Thus, after fourteen years of dispute, the Louisiana Department of Health and Hospitals (LDHH) finally approved permits for residents to install individual sewerage treatment plants on the land (Eggler 4/21/2003).

But land disputes in Lake Catherine do not end with this issue. A parcel of land called Brazilier Island was in even greater conflict during the period of data collection. Former owner Remington Oil and Gas Co. sold the land to Ken Carter, district assessor for Orleans Parish and an attorney for the oil company. When he bought the land in 2001 for an undervalued sum of $150,000, residents claimed there was an agreement with the oil company that they would get first right of refusal to buy their land if the company decided to sell. They didn't get that right. Ken Carter proposed building an upscale gated community, which would have likely damaged the wetlands that comprised most of the disputed land and would have displaced most of its residents (Jensen 2/12/2003). Prior to the storms of 2005, the dispute remained in litigation.

Due to Hurricane Katrina, however, total devastation was brought to Brazilier Island and most of Lake Catherine. At least in the near term, it may have been that developing the property appeared risky. In late September 2008 the Trust for Public Land bought Brazilier Island from Ken Carter's company for $1.1 million and then resold the land to the Bayou Savage National Wildlife Refuge. The land is considered key in diminishing storm surge and is part of hurricane protection plans by the U.S. Army Corps of Engineers. The contested residents' camp lots were not included in the sale and Remington Oil and Gas Co. still owns the mineral rights (Schleifstein 9/25/2008). Many of Lake Catherine's residents remain displaced. Rebuilding has been piecemeal with blighted properties next to large new homes built by newcomers and long-time residents alike.

CONCLUSION

These communities have their share of challenges, and their various histories continue to shape them today. Most relevant are Hurricanes Katrina and Rita. A common element that they all share is the geographic history, including Katrina and Rita, that has molded the social history of the region. Lives built around water and sediment-rich earth have fueled an attachment to place that persists in spite of problems. The biggest difficulty they all shared prior to 2005 was land loss, which threatened them with displacement. Many residents are still displaced because of the storms, and the future of their communities remains uncertain due to continual land loss and questionable levee protection, not to mention the multitude of bureaucratic and market challenges. These communities, situated in the nation's largest concentration of wetlands, cover a wide swath of southeastern Louisiana, and the great challenge they face from environmental change continues. Significantly, the attachment to coastal Louisiana that the people of these communities hold commits them to this irreplaceable region of the country, but these mounting trials makes overcoming them increasingly difficult.

4

THIS IS OUR HOME
Attachment to the Coast

We all know that certain places are special to some people. Because of the special meanings that some places hold, people often get attached, and this is certainly the case in coastal Louisiana. As in a remaining handful of locations around the U.S., many of those living along Louisiana's coast have called their place home for generations. European settlement started during the beginnings of the nation and evolved uniquely from the rest of the United States, giving it further symbolic meaning for community members to develop an attachment towards. Notwithstanding its distinctiveness, the way attachment occurred in Louisiana was the same way it develops in many places. And, as with all of those who become attached to a place, it affects their daily experience of that place.

From residents' interpretations of their experiences of land loss, we can see how the region that is Louisiana's coast is a part of how they view themselves. The attachment they have to the coast is the most important factor influencing their interpretation of land loss. For these residents, self-identifying with Louisiana's coast is due to the significant life experiences that have occurred where place both served as a backdrop and played an interactive role. As I and my colleagues have noted, this feeling of connection to a place develops "through learned perceptual practice of intimate interaction with place" (Burley et al. 2007, pp. 349–50).[1] Stated another way, place attachment is a bond of people to a physical environment based on cognition and affect which then influences behavior (Burley et al. 2007; Altman and Low 1992; Relph 1976; Tuan 1979; Proshansky, Fabian, and Kaminoff 1983). It was during interviews that residents illuminated this definition of place attachment with personal experience and revealed the role that this region played in their lives. They transformed physical landscapes into symbolic landscapes through self-definitions (Greider and Garkovich 1994). In other words, this place became socially constructed according to the meanings residents gave it.

In fact, all places are socially constructed in some way. This social construction occurs as place becomes imbued with symbolic meaning according to the values and beliefs embedded in the self-definitions of individuals. These components of identity and their concurrent values and beliefs are, in part, extracted from the larger culture (Proshansky et al. 1983; Boyer 1994; Greider and Garkovich 1994). For instance, residents of coastal Louisiana are likely to feel that they share similarities with other U.S. citizens; however, they also believe that they possess some characteristics specific to the region that makes them different from many other Americans.[2] In turn, how they see themselves affects how they see place. Coastal Louisianians see the region they live in as being like other places in the U.S., but they also see "their place" as being very unique and having a physical landscape and culture that is unlike any other. These differences of how we see places based, in part, on self-definitions should not be surprising and was evident in Freudenberg and Gramling's (1994) study, presented in chapter two, of the specifics of place that facilitated the acceptance of the oil industry in Louisiana. Again, all places are socially constructed. This does not mean that the physical elements of a locale play no role in how we see them, only that the physical has no inherent meaning beyond the symbolic meaning we decide to imbue it with.[3]

Actually, the physical context of place, the built and the natural, makes attachment possible. When individuals develop an attachment to place, first, they have significant social processes and experiences with and within a physical locale, such as someone's first camping trip to Yosemite National Park or the spending of an exciting youthful summer in Brooklyn, New York, for example.[4] These events, both in the actual event and the individual's understanding of them, are influenced by the particular social context and components of identity.[5] Next, these significant social processes and experiences that occur with and within a physical place produce sociophysical associations with place. As these personally noteworthy associative experiences accrue, attachment develops. Consequently, when we become attached to places, the physical elements become imbued with special meanings derived from experiences. Because of the special meaning that a place comes to have from our attachment to it, place becomes a part of identity. In other words, if we come to define who we are based on significant but everyday life experiences and if these events develop an attachment to a place, then our place attachment becomes a part of our self-definitions. Finally, once attachment is established it influences subsequent social processes and experiences as the physical components come to symbolize a part of identity.

NARRATIVES OF PLACE AND THE IMPORTANCE OF LAND LOSS

Coastal Louisianians conveyed this reciprocal process of experiencing and identifying with place in their interviews. As they described their communities and the coast, a major component of their symbolic landscapes was land loss. Residents broached land loss in nearly all, 100 out of 126 to be exact,[6] of the interviews, and as they spoke about the issue they communicated their attachment to the coast. The interview guide did not address the topic of coastal land loss. So it was particularly interesting that residents brought the issue up on their own. Most, like Cheyenne at the beginning of the introductory chapter, brought it up at the beginning of their interview when they were asked to talk about personal and family history. Then, many came back to the topic again and again as they addressed issues like work life, places that were important to them, their experiences with storms, and changes that have occurred to place over their lifetimes.[7]

In giving us a story about their lives and the places they live, residents wanted us to know about an important event that shapes their lives—coastal land loss. Respondents were eager to discuss the issue even when not asked. And when asked to talk about the places where they live, land loss was something respondents *wanted* to talk about. Simply, the issue had salience for residents.

In addition to their intentions, it was *how* they chose to talk about land loss that conveyed the symbolic meaning they gave to their experience. While what they said about coastal land loss is certainly important, it is how they talked about the issue that clues us in to the saliency of the event for their sense of who they are. However, I am not implying that residents were speaking indirectly and that I am uncovering something from their stories. What they said and how they chose to say it was direct. For example, most residents said that coastal land loss is ruining their community, but the fact that many chose to discuss it in the context of family history is of particular importance because it denotes the personal quality the event takes on. This choice was intentional. When any of us tell a story we are usually trying to convey something; we are trying to make a point. No less, these residents elected to talk about land loss in order to explain how important place is to them.

It was an interview, largely constructed by Dr. Jenkins, that elicited a story about themselves and place that gave residents these choices. These stories, or narratives, reveal how the "life world," as Gubrium and Holstein (1997) call it, "is produced and experienced by its members."[8] Narratives are stories that the teller believes the listener can empathize with in some way. We all do this, whether we are relaying an anecdotal narrative to our significant other about something that occurred during the day or answering the question "What was it like to grow up here?" during an interview by a sociologist. When we are interpreting our experiences, narratives involve characters that are portrayed in a particular fashion, are oriented to a type of structure (drama, tragedy, suspense, humor, etc.), and usually attempt to convey a lesson or moral (Shannahan 1999, p. 407). As Norman Denzin (1989) suggests, "every narrative contains a reason or set of justifications for its telling" (p. 41). Simply put, there is a point to telling a story.

Connecting narratives back to place attachment, narratives also say something about who the storyteller is. The point of a narrative communicates what the storyteller feels is important and by doing so reveals elements of the teller's identity. Asking residents open-ended questions allowed them to construct their own narrative. They chose how and what to respond with. By relating a narrative about place they revealed their symbolic landscapes, effectively saying, "This is what is important to me and this is how and why it is important." And, in this way, they said, "This is who I am."[9]

CONNECTIONS TO A DISAPPEARING PLACE

The language with which we choose to tell our stories is not trivial; it helps us to make our point. By using specific language, residents established this bond between themselves and place. They usually introduced this connection very early in their interviews. Employing pronouns like *we, our,* or *us* confirmed a relationship between selves and place. The two residents below exemplify the use of language in identifying with place.[10]

> Alicia (LC), a fifty-four-year-old hairdresser: We just keep losing it [the land].

> Thomas (P), a sixty-three-year-old restaurant owner: When I was a kid, I remember riding on my dad's oyster boat. We used to have two or three feet of banks on both sides of the canal and like five, six feet of mangroves. Now there's nothing out there as far as the eye can see.

In order to convey their attachment, residents often self-identified with place. Their use of these pronouns alone, however, is a mere piece of the puzzle in the larger context of meaning that constituted their high level of attachment. Whereas residents of other places use language and context to self-identify and imply attachment, these community members, in addition to verbalizing attachment in conventional ways, displayed connection through the loss of place.

> Jeppa (P), a thirty-six-year-old commercial fisherman: Like if something is missing in your house. When you get in shallow water and throw the oysters up on the boat; over the years it seems like places I keep going to, the water just keeps getting deeper and deeper. So what that tells me is that it's sinking.

> Jared (SB), a forty-five-year-old educator: And of course the swamps are retreating, and the Gulf [of Mexico] is coming towards us. So it's made us wetter. I've never had the sense of being wet like we do now. We just feel wet.

> Paul (P), a forty-two-year-old commercial fisherman: Because if the land dies, a part of us dies.

These residents purposely self-identified, not only with place, but with the effects of coastal land loss. Their attachment is obvious with metaphors of the home, being wet like the land, and death. Additionally, they equated the physical with social elements of place in conveying their attachment. For Paul, the loss of land is associated with a death of an identity and culture cultivated over multiple generations. Jeppa compares the physical loss of land with the loss of something integral to one's home, a concept that carries multiple social implications. Furthermore, Jeppa combines the social and physical in his assessment of land loss through the socially learned livelihood of harvesting oysters, while Jared appropriates the intrusion of water.

These sociophysical associations with place are something that occurred often during the narratives. Although the trend for many studies of place attachment has been to figure out whether people are more attached to the social or physical elements of place, this fusion of the social and physical appears to happen almost naturally for most of us. In other words, unless we make a conscious effort to do so, most of us do not separate the social and physical components of place. Usually, and more so it seems if we are attached to the place we are speaking of, we talk about place in ways that blur the lines between the social and physical. As was discussed earlier in the chapter, we have experiences within physical places and they then become imbued with meaning that is socially derived. Of course, places do exist independently of us, but we can only observe them in relation to ourselves. Thus, places hold relevance for us only according to the meaning we give them. As Proshansky et al. (1983) theorize, "[T]here is no physical environment that is not also a social environment, and vice-versa." Significantly, places can come to take on special meaning for us and we become attached. At this point, as Jared so bluntly illustrates, place becomes part of identity.

For these residents, attachment developed, in part, from a significant history with a particular place.[11] Natives, as well as nonnatives who lived in the region for a significant time, contextualized their discussions of land loss within narratives of family and history. During interviews, people were asked about these place-related issues and they chose to discuss land loss within that context. In this way, interviewees revealed how land loss is a critical element that impacts the whole of their lives.

> Susan (GI), a thirty-year-old graduate student: He [grandfather] taught me my first ecological lessons. You know, you save the [oyster] shells and put them back there [in the water], so the oysters have something to latch on to. It's gone. That part of the island has been eaten away so much.
>
> Lester (SB), a forty-nine-year-old commercial fisherman: It has changed 80 percent from when I was a kid growing up to now. Because like I said, the habitat is being lost; not only the seafood industry [but] the ducks and the geese. When habitat is

lost, you lose everything. This was a natural flyway for ducks and all that. And as a kid, I remember seeing ducks by the thousands, which you don't see that anymore.

Lester and Susan's passages illustrate the strong degree of attachment they have to place through a coupling of natural features with memories. Susan says, "It's gone," about the land where her grandfather taught her about sustainable ecology, and Lester takes this idea further, indicating that much more than habitat is being lost. While Susan and Lester note the interconnectedness of humans and their surrounding communities, Leroy speaks of the fulfilling experiences that were once the result of a healthy environment.

Leroy (P), a sixty-three-year-old retiree from the oil industry: I remember quite a few years ago when I first started trawling, they had a few little islands. Like during the 4th of July. We used to go out there and have a little picnic. Take the family out there. Some of my relatives from New Orleans come down, my cousin with his family. And we'd go out there and catch a few fish. We even cook out there. But as years go by, that all started washing away. They had three or four islands out there. And they used to have a lot of people out there doing that. On the 4th of July, that was a good thing to do. Get in the boat and go out there on the island and have your little picnic. It was real neat. But, you know, eventually those little islands just washed away. We used to do that. When my kids were small, we used to go out in the bayou all the time and catch some shrimp. We'd fry them out there. Take our little crab boil and go out there. My wife and my cousin's wife get out there and fry the food. Fried seafood and drinking cold beer. The kids had a good time. Play in the water until they get tired. Take a little break and then go play again. It was real nice. Mid-seventies and early eighties, we used to do a good bit of that. But after everything started washing away, we didn't have any more islands to go on.

Like Susan, Leroy reveals his attachment through special experiences that occurred with the coastal area. However, the islands that have enabled so many familial memories and cohered relationships have, as Leroy states more than once, "washed away."

Many residents' passages contain memories which include people important to them. As a result, they infuse place with symbolic associative meaning. The memories that formed their attachments included significant others and thus gave the memories that developed through an interaction with place more salience.[12] Employing these memories, residents purposely reflected upon their self-definitions in relation to natural elements in order to convey the loss they have experienced.

> Joseline (T), a forty-seven-year-old research scientist: It's obvious to us having lived here so long that the erosion of the land and the barrier islands is allowing the water to maybe move faster and come higher. And it's very unnerving. This is our home. My children are here. So it's a very serious concern to us that in years to come, I think it's going to get worse.

> Soren (T), a fifty-six-year-old research scientist:[13] But the importance of my home has started changing lately with the realization that this area is living on borrowed time. The land beneath my house will not be dry land in the not too distant future. And that affects my whole—it's hard to get attached to something that's not going to be around for long. And that's what's happening to me and my family right now.

Joseline and Soren are not natives, and because they are scientists, a more technical perspective informed their outlook. As a consequence, their attachment was more precarious than it was for most. Their tenuous characterizations of place were derived from their self-definitions of connected community members and scientists. Soren states that his attachment is in disarray and that this is "happening" to his family. Like the other passages up to this point, we can see how the environment helps to inform people about who they are.[14] Although the accounts of Joseline and Soren were less emotional than those of many others, they accordingly pointed out how their part of identity that is connected to place is threatened, causing personal, familial, and social instability.

The sense of instability that residents experienced also extended to traditional occupations that revolved around the natural elements of the coast. Residents came to define themselves through these place-bound occupations, such as the fishing industry, agriculture, or tourism, which could not be separated from other social ties to place because of generational and community overlap. These jobs constituted a livelihood, not merely work.

> Charlie (LC), a thirty-three-year-old wholesale seafood distributor: Growing up on the water really means a lot to people. You get some of these old-timers, and a lot of them grew up out here trapping. That's what they did their whole life. I've got crabbers, some of them never had a job in their life. All they did for a living was

fish. They don't know nothing else. They're really good at what they do. They catch crabs. They know where they are moving at certain times of the year, where to put their traps at, where to pick them up, and when to move them. It means a lot to those people. If this dies out, they are going to have problems. They never had another job in their life but fishing.

Charlie communicated an attachment to an occupational lifestyle that remains under threat. "Growing up on the water" displayed how this is not just a job, but rather a livelihood which became an integral part of identity formation. The relationship between the people and the land was exemplified by the expert knowledge that comes from learning this occupation as well as the "problems" that Charlie says will arise if the ecosystem "dies out." As this "dying out" becomes an increasing reality, residents of the region must consider how to act. Theodore (T), a forty-seven-year-old oil field manager and tribal council member for the Houma Indian Tribe, contemplated the dilemma.

> As a tribal leader, we met at the beginning of the year, and we did a little brainstorming through all our groups. We determined what would be important issues that the tribe would be facing in one year, three years, five years, ten years. This right here, this is the location [Theodore shows me pictures of tropical storm–ravished Isle de Jean Charles—an isolated Native American community on a finger of land at the southern end of the parish]. In the next few years, this is the track of land that we are looking at. We know they are not going to be able to stay there forever. As tribal leaders, we are looking into the future. We are going to need a place for our people living in this area to come inside the levee system. Because all of that is sacrificial land, we are going to have to invest in land. That's the only way that these guys will have a future—somewhere to go to. In the old days, the tribal leaders, they kept up with where the animals went, found new lands for them to migrate to. They had the summer land and winter land. All this had to be worked out. What we are seeing now is we are at that point in our history where we have to make a decision about which way we go from where we are at now. And obtaining land is one of the key elements that we feel that will enable our tribe to stay strong and stay together.

Theodore's thoughts drew on an idea of interdependence between past, present, future, people, community, and the natural world. He spoke of a certain amount of "sacrificing" that his people would have to do to retain strong community ties. But he also acknowledged that part of retaining community ties means retaining land ties in the same region with which they identified and maintained a relationship. He connected his tribe's situation to previous generations' symbiotic relationship with place. Theodore also communicated that for "these guys to have a future," there must be a connection to the region. Consequently, they would attempt to purchase land in the area within the major levee system called Morganza to the Gulf, which at the time of the interview was still on the drawing board, but now, some time after Katrina and Rita, has won federal approval.

In spite of Theodore's acceptance of and planning for displacement, he also acknowledged the threat to identity that coastal land loss poses. Many shared in this sense of menace to their self-definitions. Most people conveyed a degree of identity anxiety because of this threat; however, many also spoke about actions they took that served to retain place identity in the face of this disaster.

> Sylvan (GI), a sixty-three-year-old judge in New Orleans: Both my brother and my sister-in-law, his wife—she's on several committees [that deal with coastal land loss]. They're committed to this because—why? Well, they grew up and, like myself, exposed to all that. We understand its beauty and uniqueness, and we love it. It's like loving the mountain; it's like loving the desert.

Vivian responded similarly after being asked what she saw in her future.

> Vivian (GI), a fifty-year-old educator: I'm moving to Thibodaux. I'm gonna live in Thibodaux, Louisiana. It's still French speaking. They still have some French-speaking people there, and they're not going to be beachfront property for fifty years, so I'll be gone by then. I was gonna move to Lafayette, but it's too far away, I think. So Thibodaux.

Intrigued by this sense of place, the interviewer asks, "But you're not going 'til it's gone?" Vivian answers, "Oh, right. Until you can't go anymore. Well, sure! I'm the history of the island [laughter]. I have to be here." Like many others, Sylvan and Vivian directly tied their identities to place. Vivian viewed herself as an ambassador, a keeper of knowledge for Grand Isle, and therefore an integral part of the island. Her attachment is so strong and her culture is so associated with the physical region that it limits her options as to where she might move. Sylvan, on the other hand, noted the unique quality of Louisiana's coast that built a strong attachment through "exposure," a social means of interaction where someone "exposed" him to this place, thereby fusing social and physical elements. Furthermore, he simultaneously acknowledged the commonality of experience that facilitates a deep connection between people and other unique places.[15]

Accordingly, as place is thrown into question, so is identity. The activities Sylvan and Vivian spoke of—serving in coastal restoration organizations, acting as a purveyor of local history, and eventually relocating to a similar locale—are ways of actively reifying place identity. In so doing, they reconstructed their symbolic landscapes, renegotiating what place and identity meant within the context of ongoing, drastic change.[16] For most residents, their experience of land loss was mediated by what type of environment was and is still being damaged. Scholars like Stephen Kroll-Smith and Stephen Couch (1993) point out that disasters, in their social and physical relations, are affected by both the nature of the disruption and humans' understanding of them.[17] What occurs is a social and psychological process between the two elements. These residents hold a deep attachment to place which, along with the slow and sometimes quick (read storms) nature of land loss, caused them to figure out what the disaster meant to them and how to act. Like Theodore's actions of developing relocation plans for his community, the measures that Sylvan and Vivian spoke of were ways of gaining psychological control and thus mitigating the threat to definitions of the self.

Just as residents' attachment to the physicality of place developed socially, so did their experience of land loss. Most people's accounts of how they learned about land loss consisted of being educated by parents or community elders during childhood and through personal experience over time. For coastal residents, a sense of identity relative to a particular locale developed, one which was shaped by socially constructed interpretations of experiences with place. Avenues of learning such as "working with" place fueled these interpretations and led to what they felt was a type of insider knowledge about the workings of the local ecosystem.

> Rocky (P), a fifty-four-year-old commercial fisherman: All the biologists and whatever, not to put them down, scientists, but you really have to live the life on the bayou to really know what the bayou is all about. Where they had land before, I can take this boat and go right over it right now.

Liane (T), a thirty-eight-year-old community organizer: Because when we were kids, we grew up around Last Island area. We've got film when we were young and how much beach—that wasn't just a little strip; now it's nothing. And where I used to go fishing at, that land is gone. This place in Lake Pelto, we used to call it Bird Island. It's gone. How the hell could that go? It was so huge. And you go down to the island and all that, every time I go fishing, it's gone. Between Hurricane Lili and Isidore [early fall of 2002], not so much Isidore. And [Hurricane] Bill. When [Hurricane] Lili came by the game warden's camp, you could see that was land. After Lili came, you could see spots, like chunks of land just gone. When Bill came, it's open. People think I'm stupid, but I cry when I see land [gone]. And when I can catch black mullet, flounder, and sharks inside, the way I've been catching them, that's sad. Because that means more salt [water] is coming in [from the Gulf of Mexico]. More land is going to be gone. And that's how I look at it when I see things like that. When I catch fish that I know is supposed to be offshore, because I grew up fishing offshore, that's an offshore fish. [You're] not [supposed] to have that inside. And that's more salt coming in. It bothers me to see that.

Many residents expressed an intimate ecological knowledge of how the coast works, but their descriptions were also meant to communicate something more—their attachment. Insider knowledge, as Rocky states, could only come from direct and extensive experience with "the bayou." Taken even further, Liane's display of insider knowledge conveyed the drama of land loss and the emotional impact this had on her.

The attachment to place that many coastal residents have developed was constructed, in large part, out of experiences with the natural environment. Environmental psychologist Susan Clayton (2003) states that "the natural environment thus seems to provide a particularly good source of self-definition, based on an identity formed through interaction with the natural world and on self-knowledge obtained in an environmental context" (p. 51). Experiences within the natural/human context of coastal Louisiana became incorporated by the self and, in turn, informed people such as Liane and Rocky about who they are. They viewed themselves as holding a special knowledge and relationship with place that they consciously relayed throughout their narratives.

Residents utilized their insider knowledge in juxtaposition to what they viewed as outsider knowledge. Coastal community members were skeptical of outsiders who they believed claimed to possess more authoritative and credentialized knowledge. Below are the statements of a resident who occupied both of these spheres. Art (P), a fifty-two-year-old government employee in conservation services, provided an objective view of locals (insider knowledge), as well as scientists and engineers (outsider knowledge). After being asked how the scientists who implement restoration projects are received by the public, he responded:

> Well, I think sometimes they are received well, and other times they are not received well. A lot of them spend time in the communities and in the field with people in the field, and I think people respect that and they see that they are genuine, trying to research. So I think they appreciate that. And then on the other hand, sometimes they come up with findings that local people find hard to believe. They are not listening to the accumulated knowledge of generations of people that live out here. But they are forming opinions on some data they've collected that might not be relevant. So it goes both ways. But I think, by and large, people respect [the science]. I think that they know they are bright people and that research needs to be done. Finances need to be found. I think a lot of time they [residents] wish they [scientists and engineers] would listen to the local people a little more. And there may be a feeling among some scientists and academicians that they know more than the local people. And I think that's dangerous sometimes. But they are bright people, and they are doing a good job.

The disconnect between locals and those with a more technical knowledge arose often. This tension, its causes and effects, will be discussed more later. But in this instance, Art's self-definition was both local and scientific; hence the symbolic landscape he presented sought to rectify the antagonism between insiders and outsiders. Some citizens did say they were trustful of scientists, academics, and engineers, while some also showed complete distrust and opposition. However, what occurred most often was a healthy skepticism offset by a need to trust. The knowledge and attitudes coastal residents conveyed reflected their identity born from attachment. Many felt that the scientists lacked respect for their community and their "accumulated knowledge," as Art mentioned. The disrespect that residents perceived on the part of scientists diminished the likelihood that they would accept the technical knowledge of these outsiders.

Lester (SB), a forty-nine-year-old commercial fisherman: Once again when we were talking to these people, when this freshwater diversion was going [to be constructed], these are people that are light complected.[18] They wear suits. These people never had any idea about the environment that they were looking at other than what they could figure out on a computer. They had no knowledge of the area. People were telling them this wasn't going to work. But they insisted it was. I don't know how much that thing costs, a few billion. The only difference with us, we see it with our eyes. But they had like old maps. I had looked at certain bays that's in the Gulf [of Mexico]. They had one small opening to get in. You put a map against it today, there's just little strips of land, if any are still left dividing them small bays.

Tyronne (SB), a forty-year-old commercial fisherman: We went to all kinds of meetings and tried to explain it to them. 'Cause I live down here all my life. I can see the difference. It's just eating it up more and more. It looks like they don't want to hear it, I guess. I've been working in certain areas all my life since I was a little kid. I used to go with my grandpa and my daddy. Then I had my own boat since I was a teenager, and I've been doing it and I can see how much, just in the last few years since they put that [freshwater diversion in], how much it's been hurting.

Alfonse (GI), a sixty-eight-year-old retired police officer: But the engineers, they're too smart; they went to too many colleges and never come and looked at, you know, [tapping the table] not on the book, no, come and see the climate itself. Come do it. Like, not what you read out of the book. But, uh, I guess they get paid not to spend too much money.... They're spending taxpayers' money. They're smarter than me; they've got all kinds of papers to prove that they've got a degree on paper. But to me, they're—I'm not gonna say that they're dumb, but they've never been to Grand Isle, and they're gonna tell me how to protect Grand Isle?

The comments of these men were not only meant to critique "know-it-all" scientists, but to indicate their attachment, the knowledge gained from immersion in place, and to stake a claim for the value of this expertise. While sentiments such as these were widely felt by coastal residents, fishermen like Lester and Tyronne were the most vocal. The knowledge they spoke of developed from an interaction between the social and physical that created a livelihood upon which their attachment was built. Rightly so, they believed they held an intimate and specialized knowledge of the wetlands, marshes, and waterways. The fishers' heightened antagonism was most likely due to the conflation of attachment with occupation in how they viewed place. The fishing industry is the most immediately affected by restoration initiatives. For instance, river diversion projects bring freshwater from the Mississippi River into saltwater-saturated marsh areas, the hope being to restore it to its previous brackish or freshwater state and, thus, rebuild deteriorated marshland. This movement from salt to brackish or freshwater affects the numbers and types of fish and crustacean populations.

However, the fishers were keenly aware that ongoing land loss increasingly handicapped their livelihood and community. In spite of this knowledge, they were deeply distrustful of scientists and agencies charged with fixing the problem. Why would fishers not take heed of scientific knowledge and recommendations? Psychologist Glynis Breakwell (2000) offers some insight here. When a disaster occurs and a strong degree of attachment is present, identity is threatened and sensitivity is high. As a result, Breakwell notes, residents are likely to resist various social representations of their community, as well as restoration proposals they perceive to be inaccurate and coming from sources who don't share their attachment and identification with place. Of course, it may also be that in this instance the fishers, who were most affected by river diversion projects, had a financial stake in resisting, but these financial risks were not assuaged by respectful scientific personnel.

Ongoing coastal land loss threatened not only the place where community members lived and earned a living but their own identity. As community members recognized this threat to identity, anxiety built and was transmitted through emotionally charged narratives.

> Adam (T), a thirty-four-year-old port captain: It used to be a sportsman's paradise. Right now it's a horrifying nightmare. If I had to leave, I wouldn't know where to go because there's no other place; I would not want to live [anywhere] besides down here. That's why something needs to be done to try to protect it. I don't know if my kids and their kids will be able to see what we grew up in. And I hope they can find jobs out here and live their lives here like we are trying to do right now. Nothing is being done to help protect the land.... [Then, during a reply to being asked to describe his community to an outsider]: If I was them, I wouldn't even consider moving here. I would try and find another area. But where are you going to go? There's nothing. I wouldn't live up north. So I guess when this is going to fade away, I'll fade away with it.

> Susan (GI), a thirty-year-old graduate student: We shouldn't accelerate the process [of land loss by humans], and part of Louisiana's uniqueness is its seafood industry, and you need the marshes. You need Grand Isle. We are important. Just because we are a small community doesn't mean that we don't perform an important function. And to shit on us because you can—I get very upset about this.

Just like Liane, who said that she cries when she sees land disappear, these community members revealed the disruption to their identity wrought by this disaster. Responses such as these ranged from uncertainty, as when Adam expressed being lost himself and "fad[ing] away" with the coast, to Susan's anger and frustration that lent itself to a sort of lashing out at those perceived to be at fault. Not only was their attachment tangible, but their comprehension of the physical loss of land as a personal assault was also concrete.

While the loss of a place that people are connected to can cause anxiety and personal instability, survey data reveals that people rate natural settings as the most conducive for personal restorative processes.[19] Likewise, coastal residents often spoke of the advantages of living in a rural, natural region where their identities and attachments have taken on highly valued social constructions of the natural region. In other words, when people identify with and define natural places as being regenerative, the decimation of those places simultaneously erodes part of their own self-definitions.

Accompanying the damage to residents' identity was a heightened awareness of their attachment to place. If, as mentioned before, we comprehend disasters based on our understanding of the place that was damaged, then particular attachments to a place are going to shape how we view these calamities. In turn, these events have a reciprocal impact on attachment, producing statements like those from Susan and Adam. Environmental researchers Barbara Brown and Douglas Perkins (1992), studying the effects of natural disasters on attachment, found that disasters tended to produce a heightened awareness of place attachment as the perceived permanence of place was made ambiguous. Thus, disasters usually produce a drastic change to place that then produces a change in the nature of attachment. And this change in attachment is because place means something different from what it meant before the disaster. Most of these residents were keenly aware of the changes in what place meant. They actively struggled with it. Because the coast was part of how they defined themselves, changes to place meant changes to self.[20] The slow and continual loss of land threatened their sense of identity, causing a heightened awareness of attachment which they personally felt and expressed in the form of anxiety, frustration, vulnerability, and instability.

Despite the bleak outlook that most people held, they remained hopeful. This hopefulness may have stemmed from generations of community resilience in the face of natural, economic, and social hardships. Toward the end of interviews, residents were asked about their hopes and dreams for the community.

> Susan (GI), a thirty-year-old graduate student: I think—and this is the dream—I think the people of Louisiana are incredible people and I think that one day they are going to get tired of this and actually start getting together and working to see to their own interests rather than the interests of the oil companies or chemical plants.

> Chuck (T), a thirty-six-year-old commercial fisherman and oil field employee: My hopes? I'd like to see this place, something get done around here, protection-wise. Stop studying these things. Do what you need to do. Whatever it's going to take to protect us, save our land, our industry, our fishing industries. If it's not done, it's going to be all gone for us. You can dream about many things.

> Jenny (GI), a fifty-four-year-old who works in public relations: I hope that the government continues to do what it does so well right now—lobbying Congress and the state for the help to preserve us, physically, as well as our history—because when Grand Isle is gone, New Orleans will be gone. Thibodaux will be gone. And there's an awful lot of us that will lose a lot of heritage. Right now we're losing marsh. We're not losing people's homes and their families. But eventually we will if we erode away. If the marsh is all gone and there's just a thin ribbon coming to Grand Isle, that means that there'll be water in our neighborhoods and Golden Meadow [community a few miles inland] will be under water. New Orleans will definitely be under water. So [as] Grand Isle goes, so inland goes because when we're gone, they're gone.

Coastal residents' hopes existed alongside a sense of cynicism. Chuck hoped for action while expressing a sentiment we will hear many times from residents—that there was too much studying and not enough action. Susan hoped people would take back authority from large, influential industries who she believed take advantage of the land and communities. And Jenny's predictions were given credence by Hurricanes Katrina and Rita. As a natural consequence, storms wash away large areas of protective coastal land, but the damage from these storms is much more severe because of the lack of coastal land that could otherwise weaken them as they move inland. Furthermore, while Jenny believed the loss of the physical means the loss of the social, she also held faith in government, in particular her local government, to come up with a plan that would be acceptable at the local, state, and federal level.

Art, showing the pragmatism he displayed in an earlier passage, reiterates Jenny's hopes for a plan of action.

> Art (P), a fifty-two-year-old government employee in conservation services: My hopes and dreams for the area that I live in—and that goes for a lot of the surrounding areas and coastal areas—my hope and dream is the people will decide on a plan of action. I think just making a decision on what needs to be done and what they are willing to accept is going to be the biggest step. I think whatever we decide on probably can be done. Some things might not. If you're asking too much, it might not be accomplished. But if reasonable people decide on a plan of action, know what they are going in for, knowing that there might be some disruption but can accept it, if that decision can ever be made, I think we are going to be okay. My worst fear is that that decision will never be made. There will be no consensus. That's my best hope for the area.

As noted in chapter two, in 2001 a consortium of public, private, and academically affiliated scientists, engineers, and governmental agencies released a plan of action called the Louisiana Coastal Area Study (LCA) under the U.S. Army Corps of Engineers. It was deemed too expensive by the Office of Management and Budget. The Bush administration agreed to merely fund a handful of scaled-down, smaller coastal restoration projects.

The original LCA plan, although ambitious, was thought to be acceptable enough to be implemented by its architects. That plan was of the sort that Art expressed hope for in the above passage. Unfortunately, those hopes were dashed. But, in the post-Katrina era, there is special attention to coastal Louisiana's environmental degradation and thus hope regains some height with renewed plans. Project funding has increased but is still of a piecemeal fashion. Many coastal people must know, as they did at the time of these interviews, that their outlook remains a hope and not necessarily a reality. Assuredly, they are, as Sylvan (GI) says, "hopeful but skeptical."

Residents' identities are wrapped up with place in a way that is seemingly inseparable. In fact, it is what they intended for us to understand. They expressed a melding of the natural and social which has developed over generations of reciprocal interaction with and interpretation of place. They remain in the midst of a disaster which causes, as Brown and Perkins (1992) point out, an acknowledgment of previously taken-for-granted emotions concerning place. During times of relative normalcy, attachment to place resides in the background of consciousness. Postdisaster, attachment rises to the front stage of conscious thought as what has been lost or what could have been lost is considered (Brown and Perkins 1992).

Importantly, this realization of attachment at a disruptive time is a personal as well as a communal assessment of attachment at a critical moment and can be relied upon to predict how attachment might influence people when the chips are down, so to speak. It is here that attachment to place reveals much of its importance. For if, because of attachment, we change as place changes, then a multitude of reactions becomes possible because something very personal—identity—has been altered.

Louisiana's disaster of coastal land loss is slow, incessant, and foreboding. Residents live with this disaster daily, and, at the time of the interviews, it had the potential of becoming dramatic with the onset of a powerful storm.[21] As a result, this heightened awareness of attachment was never far from the foreground of residents' consciousness. Their narratives lent credence to this idea through emotion, intimacy, and conflation of identity with place in relation to the threat. That is, the threat to place produced anxiety among citizens which they then presented through their discussions of how much place meant to them.

> Sissy and Albert (Gl), married, both thirty-four years old and supermarket managers: [Sissy] Well, we've survived the storms. [Albert] We're tryin'. [Sissy] That's the only thing we think we can't be in control of, you know. [Albert] But that's what makes you appreciate it, is fearing the storms, so you know in the back of your head that, yeah, it can be wiped out. So enjoy it while you can, you know, while it's here.

> Edmund (T), a seventy-six-year-old retired mechanic: This is my home. The whole thing is important. There's nothing that's not important because everything relates to the other. If one goes, so does the other. Like I tell you, the islands went. They are trying to build them back. Before they could build them back, we paid for it in the

inside.... The only work that needs to be done is our coastline protected. That's what sustains everything down here. If the coastline goes, then everything else is in bad shape.[22]

Carmen (T), a thirty-eight-year-old office supervisor: Never take for granted that the land that you are on will always be there. Never take it for granted. It disappears in an instant. Never take for granted that you can put something in one spot and, when you come back [in] a couple of years, it will still be there.[23]

Coastal Communities

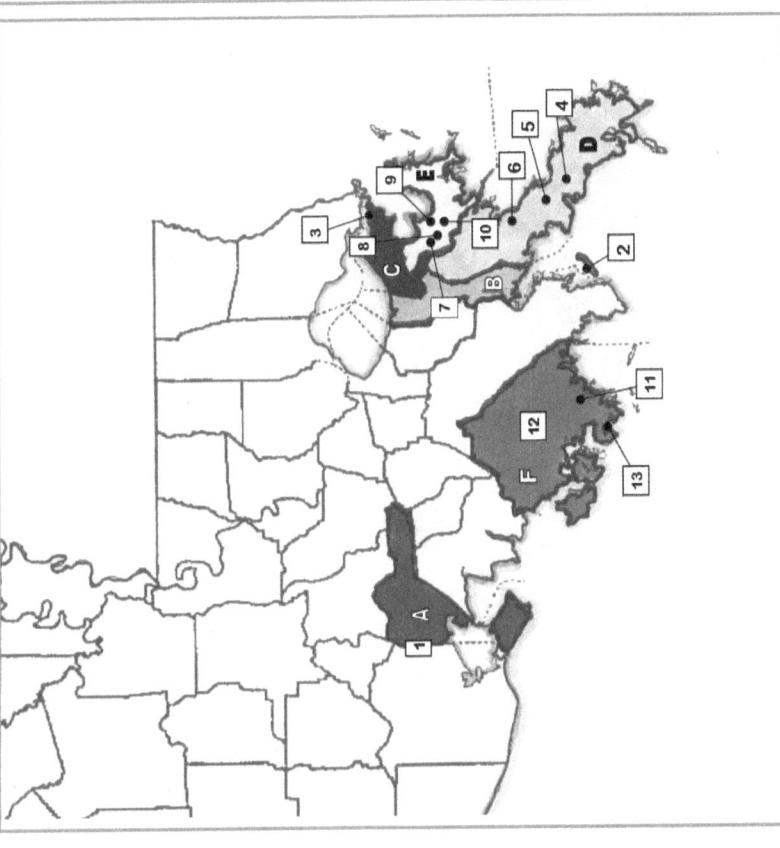

Research Parishes:

A **Iberia**
1-Delcambre

B **Jefferson**
2-Grand Isle

C **Orleans**
3-Lake Catherine

D **Plaquemines**
4-Buras
5-Empire
6-Port Sulphur

E **St. Bernard**
7-St. Bernard
8-Toca
9-Yscloskey
10-Delacroix

F **Terrebonne**
11-Cocodrie
12-Dulac
13-Chauvin

Map of the coastal communities. Original design by John Adams, 2004. Courtesy of University of New Orleans, Department of Geography. Redesign by Anne Williams, Southeastern Louisiana University, 2009.

Navigation channel amongst eroding wetlands in coastal Louisiana. Photo by Tim Caruthers, Ph.D., Integration and Application Network, University of Maryland Center for Environmental Science, November 19, 2005.

Oil and gas extraction rig amongst eroding wetlands looking towards the Gulf of Mexico, southeast of the city of Houma in Terrebonne Parish. Photo by Tim Carruthers, Ph.D., Integration and Application Network, University of Maryland Center for Environmental Science, November 19, 2005.

Intact wetlands of coastal Louisiana. Photo by Tim Carruthers, Ph.D., Integration and Application Network, University of Maryland Center for Environmental Science, November 19, 2005.

Flood protection barrier near a bayou town in coastal Louisiana. Photo by Tim Carruthers, Ph.D., Integration and Application Network, University of Maryland Center for Environmental Science, November 19, 2005.

This store, which has been in the Lapeyrouse family for generations and is now run by Cecil Lapeyrouse and his wife, is a community anchor in the town of Cocodrie. Photo by David Burley, 2003.

Middle-class professionals from Texas to Alabama—whom local residents call "sports"—build and rent camps that provide weekend and summer sport fishing excursions. Photo by Traber Davis, 2003.

Many longtime residents, like the owner of this home in Cocodrie, are leaving areas where their families have lived for generations. Photo by Traber Davis, 2003.

In southern Louisiana, politics are part of everyday life. These women advertise a rally for a local candidate for state representative who supported coastal restoration. The candidate, whose campaign slogan was "Save Our Coast," lost. After Katrina and Rita, this type of political rhetoric is an essential component of political endeavors. Photo by Traber Davis, 2003.

Early European inhabitant of Grand Isle, Louisiana. Photo courtesy of Vera Chighizola, Grand Isle, Louisiana.

Towns with large Native American populations like Isle de Jean Charles and Dulac are especially vulnerable to severe storms due in part to poverty. Nevertheless, community activist Thomas Dardar hopes that his grandson will be able to live in the region, if not the town, in which he was born. Photo by David Joseph.

This sculpture, in a garden of folk art in Chauvin, depicts a butterfly resting on an angel's palm. According to Cajun folklore, a butterfly landing in someone's hand signifies that a deceased loved one is at peace. Photo by Traber Davis, 2003.

5

SEEING IT FOR THEMSELVES

> I watched the waitress for a thousand years,
> Saw a wheel inside a wheel, heard a call within a call . . .
> —From "I Dream a Highway" by Gillian Welch and David Rawlings[1]

The attachment of residents to Louisiana's coast drives their experience of the land's disappearance. A significant portion of their self-perceptions developed either through direct interaction with place or where the coast and its communities served as a setting that helped to shape personality, as well as the surrounding social milieu. This occurs as a reciprocal process where much of the time the physical and social elements of place overlap and perceptually become fused together. This perceptual combination then informs people's thoughts, feelings, actions, and reactions to the loss, which itself is multifaceted with its political, bureaucratic, economic, natural, and personal elements. They said this experience is dynamic yet singularly unnerving.

THE DAMAGING CONSEQUENCES OF LOSS[2]

The damage that coastal land loss wrought was readily visible to residents. It was, and still is, an everyday reminder of continuing loss. While an outsider would likely only see a landscape of wetlands, bayous, and marsh, community members see the ecological ramifications of water where yesterday there was land.

> Celestine (SB), a seventy-one-year-old former homemaker and commercial fisher: I hate to see these seagulls. I hate to see them by Wal-Mart.... Because you know what's happening in the marsh. They don't have nothing to eat. They are coming further [inland]. You see these white egrets all over walking in the yards. They belong in the marsh. You see the brown pelicans in our bayous. They belong out in the marsh. They had trees where they could shelter under, and they don't have any more like that.

Roger (D), a forty-seven-year-old commercial fisher, had similar observations. After he stated that he often sees "dead vegetation washing ashore and floating in the marsh," the interviewer asks, "Do you remember that being present when you were a child?"

> Oh, yeah. It's always been. And it's just the marsh breaking up. And just vegetation that's probably years and years old. Or "flotons," which is just a piece of the marsh that breaks off and just kind of floats around. Those trawlers will catch them in their trawls sometimes. And it's a big mess. Just a big chunk of marsh that just breaks off, and it floats around, not on the top, on the bottom. That's where the term "coffee grinds," that's where it's coming from. Just a little Cajun name we got for it.

What residents notice as indicative of land loss reflects their attachment born of experience and knowledge. These images are real symbols of the loss, something that only they and biologists and geologists might notice. Celestine employed her insider knowledge of the ecosystem to explain the collision of the natural and urban. Roger used localized cultural imagery to describe pieces of land that literally float away. Sylvan (GI), a sixty-three-year-old government employee in New Orleans, paints an even more vivid picture of the problem.

> I mean the islands around the marsh is totally—every, every summer, it's substantially different. So, coastal erosion is a big, big problem. It's gonna be a big problem for the city [New Orleans]; it's a giant problem for the state and in a way for the nation because we have such a tremendous seafood [industry]. Grand Isle is probably

one of the ten best fishing spots in the world because you have tremendous bluewater fishing. You have great inside fishing. You could fish in the soil; there's life everywhere. I took my staff there, just recently, a few days ago. They saw more dolphins than they'd ever seen in their life. Birds everywhere, you know, seabirds. It's just a terrific area. And, then the shrimp that come down the estuary area from Grand Isle, Barataria, onto LaFourche and Terrebonne, it's unusual. So, it's in peril.

Sylvan, who grew up on the coast, saw the eradication of life. He viewed the region and, in particular, Grand Isle as a rich place where you can "fish in the soil" because "there is life everywhere." This is certainly an evocative account. And his point, like that of the others, is to get us to notice what is being lost. He notes the abundance of life that was then and now "in peril" and that this is not just a problem for coastal Louisiana but for New Orleans, Louisiana, and indeed the nation.

Of the various ways interviewees discussed the damaging consequences of ongoing forfeiture, the disappearance of trees was a particularly popular subject for people to use to communicate incurred harm. Gebhard, Nevers, and Billmann-Mahecha (2003) suggest that through the attachment of subjective and personal meaning to external objects, the self and object become "mentally intertwined," thus explaining how external objects contribute to identity. They go on to suggest that "the reciprocity between anthropomorphic interpretations of nature and physiomorphic interpretations of self" come together to inform notions of ourselves and natural environments (Gebhard et al. 2003, p. 105). In this way, over time humans have come to identify with trees and view them as an integral part of healthy selves and environments (Nadkarni 2008).

This cultural identification with trees is reflected by their prominence within coastal residents' symbolic landscapes.[3] The loss of trees is not only a literal clue to saltwater intrusion and land loss, but also a symbolic representation of the deprivation.

> Roger (D), a forty-seven-year-old commercial fisherman: There's no trees hardly left on them [strips of land]. It's sections that's just marsh right now. Once upon a time that was a tree line over the whole thing. It was all live oak trees. Especially from Weeks Island going back towards Avery Island towards Intracoastal City, on that section.

> Dorothy (SB), a fifty-nine-year-old project coordinator: In fact, one of the things that was noted when we had this tour several years ago for the Smithsonian [Institute of Science], we pointed out all these cypress trees with the saltwater and freshwater intrusion and all that, and how all these trees are gone. They are dead because of the fact of MRGO [Mississippi River Gulf Outlet—notorious shipping channel discussed in chapters two and three] and so forth. Really it's a sad situation because even on the property that I bought, we hate to cut trees down because you figure that tree took maybe fifty, sixty, seventy years to grow and that's a lifetime and you are going to cut it down. I moved to the country to have these wonderful things, not to cut it down. Well, the same thing holds true with all of these beautiful cypress trees that are now being killed because of these environmental disasters. They are gone. And they are hardly to be replaced. It's too bad that we can't do something like the state of Georgia does. When they cut trees, they also replant those trees. Those are the things we need to do.

Attachment to place developed with and in the midst of large, healthy oak and cypress trees. The everyday but significant experiences of place included the perception of healthy trees.[4] Whether residents climbed in them as children, fished among them, or were aware of their mere presence in yards and community, trees became part of the landscape that made up identity where the self and object became "mentally intertwined," as Gebhard and colleagues (2003) propose. While both Roger and Dorothy note the magnitude of the loss, Dorothy goes further in calling forth MRGO as the culprit and likens the fact of the dying trees to the callous act of cutting them down.

Kyle (T), a fifty-six-year-old educator, recalled the role of robust trees in his development.

> Where I'm from, Bayou DuLarge, there was a ridge of land that ran from outside of the bayou, and it went to a lake. There were huge oak trees there when I was young. And in my teens, these huge oak trees cast such cover that you were in the shade from one end of that ridge to the other. You went there, and you wouldn't see the sun except through the leaves. And by the time I was thirty, you saw the trees starting to lose leaves. The next thing you know, the trees are dying. All of a sudden, those huge trees are falling over. Everything was dying. And now there are no trees there.

Becky and William (LC), sixty-nine and seventy, both retirees: [Becky] No trees. Nothing. [William] Fishing camps on Oak River. Big live oak trees and St. Augustine grass lawn. All those oak trees are gone. There's no more oak trees. Not even the dead trees anymore. It's all dissipated. The bayous have dissipated. [Becky] It was. It was really thick like a forest. And now you go there and it's like skeletons. [William] It was wooded halfway between St. Bernard Highway to Violet Canal and Lake Borgne. It was all trees, cypress trees, oak trees, and whatever. And there are no trees there today. It's all open. If you come down the highway, you look and see some silver dead trees up there. That was from saltwater intrusion.

Trees signified and became metaphorical to the loss of land. Robert Somner (2003), environmental psychologist, and Nalini Nadkarni (2008), tree canopy biologist, theorize about the aesthetic, social, and psychological ways that trees contribute to a sense of self, and they state that our attachment to trees takes on spiritual and nearly ineffable meanings. In residents' narratives, an abundance of trees reflected an ecosystem that was healthy and vibrant, or as Sylvan said, where "life is everywhere." Likewise, the diminishment of trees evidenced the unhealthy state of the ecosystem. In this way, the demise of trees also represented the damage to the identity of individuals and community. Dying trees are a visual cue of land loss, and along with our sociocultural identification with them, it is no wonder that so many residents chose trees to express the damage of the loss.

As mentioned in the last chapter, people often report the restorative milieu of natural settings (Kaplan and Kaplan 1989; Herzog et al. 1997; Korpela et al. 2001), and the restorative effects of trees in particular are noted in numerous self-report and physiological studies (Nadkarni 2008; Somner 2003). Thus, if we identify with trees and they symbolize a healthy state, they add to our own sense of health. And when we are ill, they contribute to a sense of recovery. Accordingly, it follows that when trees die in a place with which we identify and to which we are attached, we lose that sense of healthiness of place and in some sense we feel ill. This unhealthiness of place can be a symbolic sense of illness where our self-definitions are diminished and "sick." Residents' comments in the last chapter about loss and attachment revealed this sickliness, and this is carried over here in their talk of trees when, as Kyle noted, there was a loss of a sense of protection and trees looked like, as Becky said, "skeletons" dotting the once "thick forest[ed]" landscape.

While dying trees indicated gradual land deterioration and were a continuous symbol of loss that we will hear of again and again, storms magnify and multiply that damage within a short time period. Memories of the "big" storms of the past and the seasonal and sporadic small storms served as continuing reminders that, very quickly, there could be an immense and possibly irreversible amount of destruction.[5]

> Lynda (GI), a forty-nine-year-old educator and Allie (GI), a fifty-one-year-old homemaker: [Allie] Because I think if they had something further out there, it would break the waves from coming in. Because when a storm gets in that Gulf [of Mexico], we see some terrible waves. You can just see it eating away. I mean, the levee breaks right in front of your eyes. And I think if they had something out there to break that [wave] action. . . . [Lynda] And New Orleans has to watch out. Because if we are gone, what's going to happen to New Orleans? They are going to be in deep trouble.

> Cedric (T), a fifty-two-year-old oil field employee and small businessperson: If it [Tropical Storm Bill in June of 2003] would have been a hurricane, it would have been a lot worse. But that, too, there's no more protection. If we have a major hurricane come through here and follow the path of Bill, that would be major destruction. Because there's nothing to buffer the [tidal] surge. When the tide gets two feet above normal, everything is open to the Gulf, where before it wasn't. So Bill, as small as it was, let us know it was there.

Responding to the question "Did you notice anything about these past two storms in the fall, [Hurricane] Isidore and [Tropical Storm] Lili?," Art (P), a fifty-two-year-old government employee in conservation services, explains the impacts of storms further.

> Subsidence is a factor. Everything is sinking. Coastal land loss is causing more of the water to build up. So it's striking how fast the tides come up and how high they get. Even the wind damage is pretty severe. Even when you get a glancing blow from the storm because Lili hit more the Lafayette area than us, but we had a lot of damage. A lot of high tides. A lot of erosion. We noticed after the storm when we were able to return back out into the marsh areas and bay areas that a lot of land had disappeared. Old landmarks that you used to see were gone. So it's quite evident. Every time a storm passes, you lose more and more land and you feel more vulnerable.

Anthropomorphic representations of storms are not unique. However, there was an experiential element that appeared to be special to this region. Art said that the passing of each storm built a cumulative feeling of vulnerability. This is contrary to most natural disasters where people feel a high degree of susceptibility immediately after the event that slowly recedes into a more secure sense of normalcy (Brown and Perkins 1992). Cedric buttressed Art's idea that each passing storm adds to a growing sense of insecurity by noting how smaller storms, such as Tropical Storm Bill in June of 2003, caused more damage than in the past due to increased loss of land that used to serve as protection. In turn, amplified damage from smaller storms added to the personal sense of vulnerability due to the meanings and identification residents had for place. Further, residents such as Lynda warned that their susceptibility was not unique to their community. Lynda followed the dramatic image painted by her friend Allie of "the levee break[ing] right in front of your eyes" by alerting New Orleanians to their false sense of security, a vulnerability that the city dwellers are now fully aware of. But while Lynda looked to warn New Orleanians of impending danger, community members' experience with storms also served as another source of insider knowledge, increasing their sense of intimate knowing. This knowledge was used to convey not only the destructive capabilities of storms, but also a sense of the ultimate danger that storms potentially pose due to the disappearance of land.

Cheyenne's (SB) passage at the beginning of this book echoes the statements here about the increased impact of storms. Because of its relevance, let's partially revisit her statements.

> When you build a house . . . you expect . . . you are going to pass your house down to your kids. But if they don't do something about the erosion, this will not be here in fifty years. But for some reason every storm, the water gets higher. Because that was the most water I have ever seen before I came up. And these storms didn't even hit us directly. When I built my house twenty years ago, I would have never thought that there's a chance that it's not going to be here. Not that it's going to go off, but if they don't do something about the erosion, it's going to be just water. Because that's water right there. That used to be hard land. They had pecan trees, they tell me, when my mother-in-law was a kid, like forty-five years ago. She said there was a big pecan grove. It's swamp right now. That's all the proof I need that fifty acres a year are going off in erosion.[6]

Cheyenne effectively portrayed the personal nature of the damage. She made clear that this damage is compounding over time, reiterating the comments of Cedric and Art. She spoke of the encroachment of the disaster in real terms with land turning to water and the loss of pecan trees while also conveying the impact it has had on her sense of home and well-being. In addition to the potential extinction of community from the disastrous partnership of storms and land loss, Cheyenne was also conveying a story of the possible loss of identity through the continuous thread that connects generations. The home represents a significant part of identity, and if we are attached to place, it can become a physical embodiment of that attachment. Cheyenne noted that her home, into which so much of who she is was invested, may not be available to be passed on to her children. Lending credence to that danger is what her mother-in-law told her about the grove of pecan trees that used to be across the street from Cheyenne's home. An undercurrent to her passage was a concern about the cumulative erosion of generations. When residents spoke about the increased vulnerability from storms, it seems they were speaking about more than just high water and wind.

While storms have dramatic consequences that are acutely visible, coastal land loss is much slower and can be somewhat hard to detect on a daily basis. Nevertheless, it has a tangible feel for residents. Cedric (T), a fifty-two-year-old oil field employee and small businessperson, communicated what he perceived as the obvious nature of land loss.

> You can just see it. It's so visible. It's so easy to see. You can just go out in a boat. Just ride down Highway 57 between here and Dulac and you will see. You will see the erosion going on just by the trees that are dead. And the water will start to beat up against the roads. If you have any kind of common sense, you'll know that something is happening. You don't have to have a high school or even a college education. You just go down this highway, and [you will] say, "I can't believe it was like this twenty years ago." Anybody knows that. Just look around you. You'll see it. It's so obvious. Just look across here. You can see like you can see now. The trees are all dying. So now you can see for miles across. Before, all it was was just a big old ridge of trees. You could just see the trees. That's it. Now you can see for miles out into the marsh. It's just so obvious. Anybody can see that.

Again we see the use of trees as an indicator of loss, but Cedric's perceptions were more than just noticing the visual damage of land forfeiture. The doggedly deterministic feeling of his passage implied that there was someone who wasn't "seeing it." There was a sense of desperation in his call for "seeing" the "obvious" damage before it is too late.

The following four passages echoed Cedric's sense of desperation through metaphorical language. Many residents used "death" to convey the seriousness of land depletion and the threat they believed it posed for the near future.

> Tara (LC), a forty-three-year-old homemaker: That's a shotgun looking at New Orleans. Before you didn't have that and they had a lot of filtering for that big surge to come through that marsh. You had a filter. It was like slowing that water down. But now you got a rush of water. Once that hurricane takes that topsoil, it's gone.

Beginning this set of passages, Tara commented on the danger land loss crafts by extrapolating from small coastal communities to the gruesome ramifications New Orleans faced from a powerful storm. It is noteworthy that in 2002 and 2003 many of these residents knew of the pending danger for the region. Their knowledge of land loss and what this would mean for southern Louisiana if "the big one" came speaks to their connection to and relationship with place. It is a relationship that they believed outsiders were indifferent to. The comments here from Tara, Cedric, Cheyenne, and others hold the beginnings of these perceptions, which will evolve into more direct accounts later in the chapter. But continuing with the death metaphor, Tina (T), a twenty-nine-year-old childcare employee, responds to being asked how she first learned about land loss.

> I think seeing it for myself. Going out that one time. Knowing that it took a pirogue to get through that little area in front of the camp. And now a boat [can get through]. Even in front of my mom and dad's place, that's eroding. Further up from there, there was a Mr. Ellen Duplantis; he passed away about two months ago. He had put some pilings in the front of his house. I guess with the intentions to bulkhead. And now the bulkheads are here, and the land is close to the road because he didn't do it in time. It's not just down in Chauvin, in Houma, too. They get the water. Dularge, Montegut. Everything is washing away. And even being a young teenager and hearing people saying, "There's not going to be anything left. We are going to be under water some day." And now actually seeing, someday we will. If something is not done, we are going to end up under water.

Tina chose a deceased community member to talk about his failed attempt to save his land that now deteriorates as the water seizes it in his absence. This decision was probably only chance but the symbolism is more than a little interesting. The fact that land loss represents a slow death of land and community signified by dying trees, among other depleted species, makes Tina's choice noteworthy. Almost as in a cautionary children's story, the land was neglected for far too long and by the time Mr. Duplantis tried to do anything about it, it was too late. This death, which once seemed far off in some distant future, was now knocking at the door.

> Adam (T), a thirty-four-year-old port captain: It used to be a sportsman's paradise. Right now it's a horrifying nightmare. If I had to leave, I wouldn't know where to go because there's no other place I would want to live besides down here. That's why something needs to be done to try and protect it. . . . If something is not done to protect the land, the industries are all dying in this area. But my parents had water in their home sometimes. Each time we would go to our camp after a storm, you can see the difference in the land loss. The land is sinking every day. Bayous are now huge canals. The bays and our lakes, where there used to be land, it's all open water. And in the thirty-four years that I'm on this earth, there has been a big change in this area. As far as land gone, there's nothing left. I've been up in an airplane once in my life. And that was six years ago. When I flew over the area, I couldn't believe what we were living on.

Adam referenced his strong degree of attachment, noted the morbid nature of place, said of the land, "there is nothing left," and connected this to an economic downturn where "all the industries are dying." When able to see a bird's-eye view of the area, he was in disbelief. While Adam's passage was poignant in its anxiety and sense of helplessness, the next passage employed the death metaphor in a relatively lighter fashion.

> Christian (SB), a forty-two-year-old commercial fisherman: Now there's no grass, and there's no more reefs out there; the only reefs is the man-made reefs that we have built ourselves out there which I'm sure [the] corps of engineers and stuff would never approve of us doing which I think is silly. But [the] corps of engineers also don't want us to put a load of dirt on the property. Which is silly, but that's another whole story. It [the land] has, it's decreased. This lake used to be so alive.

On the lightest side of these last four passages, Christian digressed from talking about what he felt was the frivolous bureaucracy of land loss to point out that a place that "used to be so alive" was now dead. Commonly, people displayed a sense of urgency brought on by their anxiety and a sense of helplessness. This was more than a concern for the place where they live. Their choices of how to talk about this ongoing disaster revealed a fear they had for a place that remains connected to the core of who they believe themselves to be.

> Walter (T), a fifty-one-year-old facility superintendent: I don't believe it's [the region] going to be here much longer. It's going to recede back up to Baton Rouge. They talk about this one-hundred-year event. That's it. You are not going to have anything left, the changes you see going through this marsh area which looks almost solid. You can go up in the tower and see how Swiss cheesy it is. It's full of holes. It's broken up. Between major storms, you can see a difference. You can see more open area, more open water. It makes the tide coming in and out quicker. I had asked a scientist from UNO [University of New Orleans] one time. The locals kept talking about the higher tides look like they are coming higher and the lower tides look like they are lower in the wintertime. He said that's probably true because of your barriers. You don't have the barriers to slow that tide down. So it's an easy flush up and down. A lot of people are moving dirt into their property down here, trying to elevate it, keep it dry. What happens is once it becomes wet for the majority of the year, you start getting marsh grass growing. It's a little harder to cut with your push mower.

Walter's statements, like the tone of his entire narrative, were somewhat rational and objective. Echoing the sentiments of his neighbors, he indicated the area's sense of vulnerability by noting the lack of protection. He then employed a conversation that he had with an "expert" to buttress locals' perceptions of the increasing degree of land loss. While he was not as emotional as some others, rational fatalism such as Walter's was not uncommon. Importantly, Hurricanes Katrina and Rita may have enhanced the beliefs of residents like Walter that the coastal line won't just recede to New Orleans, but eighty miles north, all the way to Baton Rouge.

No matter personal nuances, all of the passages here communicate the *damaging consequences* that people saw resulting from coastal land loss. The meaning that these community members conveyed in this theme was that the damage to place was also damage to identity. There was the loss of a nurturing ecosystem that touched everything from fisheries and birds to trees, land, and residents' well-being. The self-definitions of residents held a shared identification with place that took on meanings associated with the damage their communities were experiencing. As a result, they felt desperate, unprotected, and anxious. In a way, it seems that the refrain of "If something isn't done . . ." that we heard many times and that we will hear many more was a plea for the salvation of themselves. The harmful consequences were not only to the land or even to their livelihoods, but to who they were.

RESTORATION

Everybody has to build up their ground.
—Claude (T)

When residents talked about land loss, restoration was the theme that came up third most frequently. Citizens discussed how the damage could be stopped or at least diminished. Many held conflicting and ambivalent views about the restoration process, but everyone urgently wished something would be done. And while almost all heeded the burning nature of the problem, some saw the issue of rehabilitating the coast negatively and some positively, while most fell in between. But again, their meanings were contextualized within an attachment to land, community, and the region.

> PJ (GI), a sixty-four-year-old former oilfield supervisor: They [governmental agencies] put rocks and they pumped in the beach and some areas. Industry has bulkheaded, filled in the back [north bay of the island]. The [U.S.] Coast Guard put rocks all the way around, and all [along] that end of the island is built up and the Caminada Pass side, they've done the same. They put rocks and sand and shrubbery and built that in. We are in a lot better shape now than we were in the early fifties.

And Conrad (D), a forty-nine-year-old educator, responded to being asked if he had kept up with any of the state's restoration proposals.

> A little bit. I can see, not far from up there, they are damming it up with rocks, both sides of the Intracoastal [Canal]. It dammed up the mouth of the Avery Canal which we call the Delta Canal. It's not as wide as it used to be. You can only get one fishing

boat through there. I know they're doing some work out at the Coastal Canal, and I think they are planning on doing some work on part of the Oaks Canal.

In both passages these men framed restoration in a positive light and connected themselves to place through a display of insider knowledge. This was more than a cursory grasp of what was going on in one's backyard, even for Conrad, who said he had kept up "a little bit." Their knowledge about restoration efforts was contextualized within their everyday experience of place and the processes that occurred there. As a result, these citizens were fueled with the sense of someone who has a specialized proficiency of their setting. Conrad spoke of land being built along different navigation canals and how one "can only get one fishing boat through there" as opposed to many others who said canals widened over time where there had been no restoration attempts. PJ similarly displayed his localized knowledge and self-identified with this recovery process, saying that "we are in a lot better shape than we were." Residents like PJ and Conrad saw some recovery of place, thus implying the illness of the region that was evident in the *damaging consequences* theme.

Although some said positive things about the restoration process, many people possessed negative perceptions of these efforts. Even those such as Conrad and PJ who had favorable words about projects were also critical of restoration issues. Tyronne (SB), a forty-year-old commercial fisherman, speaks of the polluting effects of good intentions.

> And they got that canal right in front of the house. You can go fishing, and we've been riding in boats a little. Before they put that siphon in Caenarvon; they used to have a lot of clear salt water used to come in. You used to see dolphins in the bayou in front of the house. We used to go water skiing. We did a lot of swimming and skiing in the summer right there in the canal. But now they got—I don't feel the water is safe no more since they put that siphon in. They got the river draining in here, and that's polluted. You can feel it when you get in the water. We get in the water to work on our boats to change a propeller or something. We jump over and do it ourselves. And it actually burns your skin. You can feel it.

Tyronne is then asked if he knew when the siphon project was implemented.

> I don't know exactly the date. It must be at least seven or eight years. They run it sometimes. Then they cut it off certain times of the year and stuff like that. I don't

> think it's really helped it [the land]. I think it's really hurting. That water ain't too good. We get in it all the time. I got a swimming pool in the backyard for my kids. I don't want to swim in that. And we grew up in that. We were raised in that. It was all salt water. You can see a difference. The dolphins used to come in the canal. You don't never see nothing like that no more. They not going to come in that river water.

Here, Tyronne critiqued one of the major restoration projects, the Caenarvon Freshwater Diversion Project designed to divert freshwater from the Mississippi River into the marshes of St. Bernard in order to rebuild land and restore a brackish water state to what remains primarily salt water. Tyronne implied that the canals and bayous were naturally salt water while he was growing up. Although the saltwater state he spoke of was not natural but a result of land loss, he juxtaposed the previous saltwater condition of his community with the polluted intrusion of a failed human endeavor to improve the land. The southern Louisiana section of the Mississippi River is widely known as a polluted waterway due to runoff from the oil, gas, and chemical plants that line a section of the river notoriously known as the "chemical corridor" or "cancer alley." This pollution is compounded by chemical agricultural runoff from Midwest tributaries that drain into the great river. Incorporating this commonly held community knowledge, Tyronne communicated that it was not just that the Caenarvon Project wasn't working; it was further damaging his environment. He symbolized the damage by the "burning" of the water and the lack of both fish and dolphins, popularly and positively anthropomorphized mammals. He added further weight to his criticism of the restoration project by stating that the waterways of his home were something he needed to protect his family from.

Next, Jeppa (P), a thirty-six-year-old fisherman, continues the critique of restoration projects while also pointing out the absence of community participation in these endeavors.

> I hope that we can really get some people in government that really knows, or try to at least figure out, how to save our coast. They need to get with the people who work out there and live out there to really get a feeling on what's really happening. As far as the freshwater diversion, I think that's a waste of money. I don't know how many millions and billions of dollars we've spent. Why not just build the land back up. Get some old maps. Just build it up. I think that would be the best thing because that's like a rebel with a sword, too. Who knows what the future holds?

Jeppa echoed Tyronne's aversion to freshwater diversion projects while pointing out the gulf between those who would fix the problem and those who live the problem. While coastal geologists agree that the best way to restore Louisiana's wetlands is through river diversions and sediment delivery systems using pipelines, it was obvious that many community members were not won over. Tara (LC), a forty-three-year-old homemaker, was asked if she thought restoring the coast was possible.

> No. I don't know. It wouldn't be a hard thing to do. Because if you shrimp and crab here, what's the difference of going out to the open water? But I'd like to see somebody do it, though. But they're not. They just gonna let it go and go and go until it's too late again to do anything. And it may be now; I don't know. I don't know how the—I know that the Lake Pontchartrain Foundation has built a few man-made reefs with the Christmas trees and stuff. I don't know how the progress is.

Tara was more ambivalent than Jeppa and Tyronne. She took note of a piecemeal measure where discarded Christmas trees were gathered and used to build land, while she also viewed government and its related agencies as dragging their feet. Tara believed no significant action had been taken, and that it may have already been too late.

> Robert (GI), a sixty-one-year-old retired educator: In fact, back in the fifties the government came by and said, "We're going to help you." And they were going to improve the property. They were gonna pump the channel in the back of the island, and the mud they were going to pump out of there, pump into the marsh area and make it high ground, so they put a levee up and they put it all up and of course it destroyed the marsh.

Robert reiterated the negative consequences of good intentions expressed in the three passages prior to his. He pointed out that the attempts to protect a community from flooding also "destroyed the marsh." Interestingly, as residents show below and then later discuss in the *politicized* theme, it seems that the alienation these citizens felt—where outsiders inflicted projects on an undervalued and disrespected public—may be a contributor to restoration aversion.

While some residents expressed disdain for agencies charged with coastal restoration, sacrifice was the meaning that the next two people intended to convey. Most community members, even those discouraged by existing efforts, acknowledged that some sort of capitulation on the part of communities will be in order if the coast is to be saved. Currently, commercial fishers appear to be sacrificing the most as restorative diversion projects change the current state of the ecosystem in ways such as transforming saltwater or brackish waterways into freshwater systems. Nevertheless, some fishers don't believe there should be any sacrificing on their part. Recall the oyster bed dispute discussed under the Plaquemines Parish history section in chapter three. In the mid-1990s, a controversial lawsuit was filed and originally won by oyster fishers of Plaquemines Parish against the state of Louisiana. Because of what many saw as egregious monetary awards, this case threatened the feasibility of future funding for regenerating the coast. Louisiana appealed and eventually won against the original decision in favor of the oyster fishers.

These next two residents refer to the infamous lawsuit. Larry (GI), a thirty-six-year-old restaurant owner, begins by explaining the Davis Pond Diversion Project.

> Davis Pond Diversion is a siphon that they built on the [Mississippi] river and it's around Avondale. And it's on the north side of the Barataria Estuary. And it's going to just flood a lot of freshwater and sediment from the river into the upper Barataria Estuary. And I don't think that one's going to be enough. I think it needs more. I know you start affecting people's livelihoods—oyster fishermen, things like that—but I mean, I think you got to look at the big picture. You know, if it keeps eroding away there won't be any oyster fishermen anyway. So I mean, if they got to move further south—you know, fifty, sixty, eighty years ago they weren't fishing oysters there anyway. So I mean, I think that's the only way they're going to be able to do it, you know. Right now between Grand Terre and Lafitte you're not talking about a whole bunch of land, and there's hardly any left. When Barataria Island is gone and a few more of these little small islands on the north side—ten, eleven miles from here—there's nothing left. And I don't know what they're going to do. I think they need more siphons, you know.

Saro (SB), a fifty-seven-year-old land surveyor, responds similarly after he brought up the Caernarvon Project and was then asked about possible problems with the diversion.

Well, the thing is that the fishing industry in St. Bernard Parish has developed around a deteriorating wetland. And their practices, their equipment, the areas that they lease for oyster bedding have evolved into a process that really depends upon that deterioration. Now we come here and we say, "Well, we are going to stop the deterioration." We are going to reverse the process when all the practices, equipment, leased areas, and all that become obsolete. And they [fishers] don't like that. They want to be able to continue practicing the fisheries as they always have for as long as they can remember, which is two generations. And that's about as far back as the practice goes. The sad thing is if we continue to allow it to deteriorate, they are going to find out they are not going to have any fisheries at all. That's going to be many years down the road. So how are you going to tell a guy today that's making money, say in oyster fishing, we've got to disrupt your business so that fifty years from now, a hundred years from now, they will still be able to do oyster fishing. Well, he'll say, "Fifty years from now, I'm going to be comfortably dead. So I don't care about that. I would prefer to have my money and make it now." So it's a natural and understandable human reaction.

As Larry and Saro saw it, the condition of the inland bayous and canals was anything but natural, as Tyronne implied earlier. Larry referred to the Plaquemines oysters lawsuit and suggested that an inability to sacrifice would exacerbate an already urgent land situation. Saro also echoed the idea of necessary sacrifice but was more empathetic about the fishers by noticing how economic forces propel people to act in a self-serving manner. While not indicative of a willingness on the part of fishers to make sacrifices, the Louisiana Supreme Court ruling to repeal the Plaquemines oyster fishers' awards did reflect popular sentiment among residents. While most believed that people should be compensated in some way for lost income, they also believed that too many had unfairly tried to profit from the lawsuit.

During the era of that dispute, before the storms of 2005, commercial fishers in Louisiana were in the midst of tough economic times. With regulations, the increased cost of insurance, fuel and boat maintenance, and competition from aquaculture imports driving down the price distributors would pay for a catch, many claimed it was harder than ever to sustain a living. Consequently, since the agencies that were in charge of coastal restoration were viewed by many fishers as yet another obstacle, some of their frustration was projected onto the restorative process. Rocky (P), a fifty-four-year-old fisherman, begins the next two passages by explaining his thoughts on the extent of scientists' and biologists' consultation with fishers.

> Yes. They have meetings. They come meet all the time down here, but they don't listen. So why not go where the problem started, where the land started to wash away years ago? Build it back up there.

> Adam (T), a thirty-four-year-old port captain: They [agencies] see what's happening. We all try to voice our opinions on things that can be done to help save it. But we don't see anything happening. Nothing is being done. They tried different things, but it doesn't seem to work. And all that is another drop in the hat when it comes to things like that. We can voice our opinion but nothing gets done about it. They do what they want.

Rocky and Adam expressed a common sentiment among residents. Whether it was the Louisiana Department of Natural Resources, the Louisiana Department of Wildlife and Fisheries, scientists with the USGS, or the U.S. Army Corps of Engineers, people believed that most "don't listen." While fishers were the most vocal, community members of all occupations felt dismissed and that their knowledge was insignificant relative to the authority of science.

> Lester (SB), a forty-nine-year-old commercial fisher: Nobody ever went out to the person that does this [fishing] on a daily basis and said, "What do you think when the wind blows this way, what's the best way to go?" Nobody did that. They all said they had statistics and all this scientific findings. But it's like anything, scientific findings. They are confined to a four-by-four tank in a lighted room out of the elements, and they give you the outcome. Where you take somebody that's not a lab rat and give them the real run.

Lester expressed the marginalization of local knowledge that many felt, and his point was clear: the collective knowledge of those in his livelihood has no value within the institutions of power. Lester acted back against this perceived slight by pointing out the detachment of those studying the issue. In doing so, he indicated the daily connection of himself and comrades, people who have, in many cases for generations, lived, worked, and gained an intimate knowledge of the places that are turning to water. Next, the sentiments of Paul (P), a forty-two-year-old commercial fisher, about the disregard of his community appear to be particularly incisive.

> And what's actually happening is they are going bigger and grander. What used to be seven billion is now fourteen billion [dollars]. It's this big grand push to get the huge federal dollars. I don't like the attitudes and the philosophy of the institution of coastal restoration at all. Because at first, we went to them looking for support. Then something organizes. And then it grows into this big permitting process and this big arena of studies and millions of dollars being wasted on these studies. When dang, just go do something. Quit studying! It's a constant battle. And then to be used to drum up their support, to be federally funded to support their little network. And also this little network is turning into this big conglomerate. Now they're stepping on us. That's kind of how it feels. I don't have an active voice in it anymore. I should go [to meetings], and I still do what I can with what time I have to give trying to earn my living. I don't get paid [in giving time to restoration issues]. Most of them people are paid because that's their job. And they sit around talking about it all day. I'd rather pick up a shovel and carry some sand on my back and do something, so to say. And it's just frustrating when you are trying to present this philosophy, this project and you're standing before these guys who have never been out there. And they kill it, for whatever reason. I can understand their position to have limited amount of money and stuff like that. But they just don't see; we see the importance.

Frustration and alienation resonate throughout these last several passages. Since these residents understood coastal restoration as a divisive process, it is not surprising that many projects were and still are met with community opposition. Consequently, communication between agencies and communities spirals into conflict. The result is far less than what should be done to adequately rehabilitate Louisiana's coast. It seems Paul expressed his resentfulness and disaffection the most poignantly. In the past, he gave much time to cooperating with scientists studying coastal land loss, even continuing to maintain a close friendship with a local university researcher. However, he believed it was no longer about community cooperation. He thought that "they're stepping on us" and that he and his community were only "used to drum up support" for self-serving agency projects.

It doesn't have to be this way. It is possible to change the paradigm of conflict between community and bureaucratic organizations. As will be discussed more thoroughly in the concluding chapter, changing this conflictual relationship is necessary for sustainable restoration not just of Louisiana's coast but for environmental reconditioning elsewhere as well.

In fact, this shift in how restoring the coast is carried out will be necessary because of the attachment that residents have to their communities and the region. The strong attachment residents had to place was where their resentment and alienation emanated from. They considered themselves a part of place yet perceived being shut out of its recovery process. On the other hand, one might ask, "Why don't the people of these communities just let the agencies implement restoration projects?" One could even use residents' own claims against them—"Yes, the scientists in agencies such as the U.S. Army Corps of Engineers sit around studying this problem all day. They know how to fix it. Let them." Yet, claims such as these miss the strong sense of identification and possession residents have to coastal Louisiana. Community members' perceptions of the agencies' denial of their expertise is akin to authoritative rejection of personal knowledge of all sorts from a physician's dismissal of what we are experiencing with our own body to someone's repudiation of our personal feelings. Furthermore, considering the damage to the ecosystem and residents' identification with that damage, it becomes easier to understand that not being allowed to engage in the restoration process led them to think they were shut out of recovery. In spite of this obstruction, or perhaps because of it, some residents spoke of taking remediation efforts into their own hands.

> Jackie (LC), a fifty-one-year-old fireman: I don't think diverting the river is going to do anything to stop the influx of water from the Gulf [of Mexico] onto the land. That's what needs to be done. You have to build the land back up. And I don't think the river is going to put that much silt to build that land up quick enough to help anything. . . . If it wasn't for the people out here who do fill in and put mud in and put "wash out,"[7] if they wouldn't do that to the land out here, this island wouldn't be here anymore. It would be washed off. But it's from the people living out here who have taken care of the land and doing this and that, to improve that land and build it back up. . . . I think it's up to each individual landowner to do it. I don't think the groups have anything to do [with] it. I don't think the groups have the control of what an individual person does to our land. It's up to that person to put filling up or bricks or rocks or whatever they may decide once it starts eroding. Because you may have somebody who owns a piece of property, and I've met quite a few

of them. If they wouldn't have put rocks or bricks or something down years ago when they were living there, now there's some people that's dead and gone off this island. Because they did it [filled in land] in the past, the land will still be there.

Adam (T), a thirty-four-year-old port captain: But you can see the difference in the land. [Adam is showing the interviewer family photos of their camp dating back three generations.] And then in '93 we started building this one here. We finished in '94. Right now the water level is even with the marsh. What we do now is put a bulkhead along the bank, and we haul oyster shells and rocks from our house out there to try and protect what's left. It's helping but you can see in the background it's eroding from the back side now. So we are going to have to soon start doing something back there unless another storm would take the camp.

Part of the meanings that residents commonly attached to coastal restoration was that of their own agency. Jackie, who continued the thread of skepticism of institutionalized projects, symbolized the reconditioning of land through individual and community responsibility and action. In their research, Austin and Kaplan (2003) have shown how actions that restore one's damaged habitat also serve to repair a part of identity injured by environmental damage. In this way, Jackie also renegotiated his own self-definition regarding place as well as that of community. Accentuating the notion of group action through individual efforts, Jackie even credited land preservation and the continuation of place to those who are deceased. Likewise, Adam noted his family's battle to save their land while reifying his attachment to the coast through displaying photos.

Kyle (T), a fifty-six-year-old educator: I went into education and really got into a lot of community things, environmental things. Right now I serve on the Coastal Zone Restoration Committee for this parish. And other organizations that try to stay on [top of] what's going on—familiar with what we are facing. . . . I believe something will be done [to save the coast]. Now there's a lot of projects. Now the big topic, every time you pick up the paper, the politicians are talking about doing something about it. And there's a lot on the national level now. As we are talking to get federal dollars, we have taxed ourselves. The taxes passed a couple of years ago to help match local dollars needed to get the federal dollars. And no sooner than those dollars are already accumulating. The next thing you see on national news as these bills are starting to come up, CBS, CNN are all telling stories about coastal settling.

Kyle, unlike Jackie and Adam, was involved more formally with coastal restoration. His belief in community agency transferred into his hopeful meanings and intentions. While Jackie and Adam also showed faith in community, Kyle believed that democratic legislation in the form of self-taxation would show commitment by his community and thus lead to exposure at the national level and stronger federal commitment. And while Kyle was committed to more prescribed modes of restoration, Gerry (T), a forty-six-year-old educator, took a very hands-off approach. After discussing land loss and the unfortunate future he believed southern Louisiana faced, he was asked for his thoughts on the public dialogue by the state and other efforts for restoring the coast.

> It's all talk. I just look at it as being all talk as long as the oil companies are there and those refineries and those companies are still alive to be out there. You've got to leave the earth alone for a while to let it heal on its own. And as long as those few people are still out there cutting and butchering and dumping, it's not going to heal. They do talk a good talk. I've always seen one attempt that looks like half-way decent work. It was involved with Save the Lake Foundation, saving Lake Pontchartrain. And I remember when there were no pelicans in the lake. I helped work in putting lake grass back in, introducing cypress trees back in a certain marsh. And I was at the museum in Kenner [town in metropolitan New Orleans area] one day and I was doing a lecture and I stopped dead in my tracks. And this was about eight years ago. I looked up and there was an endless line of pelicans coming from the coast into Lake Pontchartrain. After I got 250, I just stopped counting. And I went, "My God, if they leave it alone, it will heal." And that's what they did. They left the lake alone. And the only way they are going to do that is to do the same thing. Leave it alone. Stop dumping crap into it and stop these companies from doing the stuff that they are doing and control the amount of whatever goes into that lake. You are going to have it done. Man is not going to do it. Man will do it if he leaves it alone. The earth will take care of it itself. You are going to lose a lot of coastline. But I think the way the earth moves is that you lose some, but I think the rivers will reroute, the water will work, and eventually it will get to a point where it's going to stop. A lot of times man thinks he can do anything. You can't. I mentioned one day, I would love to see if everybody pulled out of New Orleans for ten years and you never came back in those ten years. You could walk in, there would be a rainforest, because the earth would take it over. That's all they've got to do. Leave it alone. I think a lot of it is lip service.

Gerry's symbolic landscape of the coast was unique. Nonetheless, elements of how he defined coastal land loss were common. Like other people, he blamed powerful economic actors for the deterioration of the region, and he expressed a wish to see the land and community preserved while noting a skepticism that it will actually occur. Additionally, he thought that mainstream cultural norms have produced a hubris leading humans to think and act toward nature as a passive, submissive object—humans can harm their environment while they can also, by force of will, make it well again. However, Gerry's desire for the absence of human action upon the land was different from most residents' beliefs about restoration. He believed that the environment has its own agency, that it can recover and thrive independently of humans. He buttressed his theory with an example of ecosystem self-revitalization that gave credence to the meanings he applied to the coast.

Although no other community members expressed the depth that Gerry did about restoration or ecosystems, many did allude to this sort of belief system. Many stated that human selfishness has produced the current degraded conditions, that society must cease all negative actions upon their environments, and that restoration may be only empty rhetoric. Gerry's statements and those similar to his didn't develop because the speakers had some special or mythical connection to "the land." These community members have watched a place that they hold an attachment to steadily deteriorate. And like others in many other places, they have the collective memory, the intimate familiarity with place that was passed down to subsequent generations, to recall what their healthy state was like.

On the other hand, this collective memory is unfamiliar to many of us who move to someplace new and can't remember what it used to be like or those who are only aware of the changes within their own lifetimes. Environmental psychologist Peter H. Kahn, Jr., calls this environmental generational amnesia (2002). Interestingly, earlier Saro indicated that some fishers might have had this special type of amnesia when he said that they couldn't, or didn't want to, remember that the marine life they harvested previously did not prosper in the areas where they now find them. Notwithstanding some, Gerry and those who expressed beliefs similar to his didn't seem to be afflicted by this condition. They developed their beliefs from self-definitions that comprise a social and natural history that are not separate but overlap.

While Gerry's affinity for the natural aspects of the coast was evident, in other parts of his narrative he revealed his detachment from the region. He had not lived along the coast for some time. Considering his strong feelings about what restoration means and the almost impossible reality of such an idea system, it was no wonder that he was detached. On the other hand, for those who only alluded to such sentiments and operated within mainstream cultural schemas of how humans should interact with their natural systems, it was also no wonder that they expressed a sense of urgency about restoring the region.

Rachelle (T), a sixty-one-year-old chef, responds to being asked if she thinks anything can be done about land loss.

> Probably. If they would start doing something with their surveys instead of doing another survey. I'm all for surveys. Don't get me wrong. But once you find out what you need to do, get out there and do it. Don't drag your feet until you need another survey. It's just wasting money and time, valuable time.

On the topic of time, Alfonse (T), a sixty-five-year-old retiree, believed that some communities were already out of that precious resource and that some displacement was inevitable.

> You see where we are at now, we are inside the levee. Pretty soon it's going to be like New Orleans. Every time it rains you get water. But I believe that something needs to be done to protect what they got left. I wouldn't worry about all those islands now. That's too far gone.

Next, Tina (T), a twenty-nine-year-old child care employee, ratchets up the sense of urgency.

> We see that where we are now. You bulkhead and you bring in oyster shells to try to save the land around the camp. But you feel like you are fighting a losing battle. We bring cement out there, oyster shells by the baskets. It seems like you are not winning. . . . But if people don't start doing something about the wetlands, people need to stop talking about it and just do something. They do have little programs out here where you can go out and clean up. They go out to Last Island and clean up in other areas around here. But we need to do something.

Residents have been experiencing coastal land loss for generations, but it is only in the past fifteen to twenty years that it has begun to enter the popular consciousness of the larger region. It was common to hear that there is too much "studying" of the issue and too little action. Everyone had opinions about how restoration should proceed, and, considering the alienation that people felt from the institution of coastal restoration, it is not surprising that many were skeptical of the efficiency and reliability of projects and what seemed like a never-ending stream of well-funded studies. Alienation from institutional processes and the futility of individual mitigation efforts added to a sense of urgency within residents' conceptions of place. As citizens watched their places disappear, their experience of coastal land loss took on added meanings of anxiety, as expressed in Tina's statement when she said that we need to "stop talking about it and just do something."

When residents spoke about coastal restoration they conveyed positive and negative meanings that arose from their experience of rehabilitative procedures. But their interpretations did not develop singularly. Their perceptions of redevelopment processes emerged within the larger context of land loss together with their attachment to community and the region. For these residents restoration involved sacrifice and conflict. Saving Louisiana's coast took place within a framework where those who have an intimate relationship with place felt dismissed, undervalued, shut out, alienated, and distrustful towards those charged with alleviating the loss. The restorative and mitigating actions that individuals and fellow community members engaged in served to offset these negative feelings. Individual efforts helped them regain some sense of autonomy and reestablish a sense of self in relation to place that was damaged not only by the disaster of coastal land loss but from the human process of coastal restoration. These self-restorative actions aided residents in dealing with their sense of helplessness in spite of the fact that they held humans and even themselves responsible for the ongoing disaster.

HUMAN DEGRADATION OF THE COAST[8]

As has been indicated throughout, residents held humans responsible for a large part of coastal land loss. Although the other themes have obvious references to how people apply responsibility, it is in this section that their clear intention to confer culpability on the actions of humans is given focus.

Most community members talked about the larger, well-known human causes of land loss—oil exploration, the levees around the Mississippi River, and so forth—but many also gave more localized meaning to the human influences of the disaster. Alysha (GI), a forty-six-year-old librarian, said that much of the erosion on Grand Isle was inevitable due to natural wave action. She went on to say:

> Another thing I'm always on people about, you're only supposed to go across [to] the beach, across the levee, on the boardwalks. And people will build a camp and then they'll go cut down the grass and then they'll make their own little crossover and they're not supposed to do that. And they're not supposed to drive four-wheelers and golf carts and stuff like that across the levee. If there's no plants on the levee holding the sand together, that's the first place it's going to erode, and the people don't understand that. Of course, it's just weekend people and their camps are insured so they usually don't care. The people who live here know.

Alysha, who spent her childhood in the Midwest, viewed herself as an attached insider who was a steward of place attempting to alert outsiders to their detrimental actions. As was common among residents, Alysha displayed an identification with place where a self-perception of stewardship was juxtaposed against that of uncaring outsiders. Let me reiterate a function that these passages served for residents: by intentionally conveying this understanding to the researcher and verbalizing her actions, Alysha reified her identification with and attachment to place, even self-fortifying it.

As a component of self-identifying with their environment, residents viewed themselves as insiders with a certain amount of environmental expertise. As researchers Opotow and Brook (2003) have shown, community members tend to perceive government agencies and special interest groups like environmentalists and developers as outsiders who are disrespectful and dictatorial. In addition, most of these communities were undergoing an influx of upscale suburban development that intuitively went against coastal restoration. Development of these areas requires wetlands to be filled in for building, an issue which Phyllis (SB), a forty-seven-year-old office administrator, tackles here.[9]

> Just like the houses [development], these people that come in and knock the trees down, to me that's another thing that takes away from the land.... At one time some of these things around here were like instead of wetlands, and all of a sudden, now you see some areas developing in and around those areas. So it's like a lot of the trees have died.... Some of the trees have died because, well, they either

have to remove them or they have to fill in where these trees were.... They'll say from this point on, this is where [logging stops]. The common people could never do nothing with them. But then again, I'm not going to mention names, but you get certain types of people that purchase this property. All of a sudden, the lines [of demarcation] move either across to the side or they move them back. And then this land that no one never could do nothing with before, it develops. Figure that one out.

Again, trees came to symbolize the degradation. Phyllis viewed powerful economic actors as bending government to their own will, whereas "common people" were restricted from using the land that she thought was their right. Alicia (LC), a fifty-four-year-old hairdresser, echoed this insider status, but where Phyllis's thoughts about conservation were unclear, Alicia made a point of her stance on the matter. In this passage Alicia was in the midst of discussing her conversation with someone in city government in charge of zoning, planning, and development.

He said, "First of all, you are the first person I've ever spoke to that didn't want to change our wetlands. Everybody that calls me wants to find out how they can get around [regulations] to be able to develop wetlands, or to use it." He said, "It's like a breath of fresh air."

Alicia positioned herself as a protective custodian and others as manipulative degraders of the wetlands that she identifies with. These meanings of insider/caretaker and outsider/degrader added further to residents' constructs of coastal restoration as a dishonest process where those with power could circumvent the rules to the detriment of the land and the people who live there. Human degradation of this sort also informed symbolic landscapes of coastal loss through the simultaneous representation of damage and disrespect to the land and identity. To put it another way, the perception of underhanded degradation for profit further damaged a land already "in peril," as Sylvan said earlier, and added further insult to those who are deeply connected to the region, making them feel as if they are "shit on," as Susan said in chapter four.

As development symbolized the exploitation of place by outsiders, so, too, did the oil and gas industry as it came to represent the chief environmental degraders in the popular consciousness.

Christopher (LC), thirty-eight-year-old small businessperson and former fisherman: The oil rig used to be three hundred feet in the marsh. Now it's sitting out in the middle of the open water. There's no canals through there to take outlying canals. I'm sure you heard that one. They've got all these pipeline canals through the marsh. All that did was cut it [the marsh] up. It gives access to small boats to run even faster. So believe it or not, it should all fall back onto the oil company responsible for it.

Becky and William (LC), sixty-nine and seventy, both retirees: [Becky] I think the oil companies ought to put something towards that repair. [William] They let all of that salt water in, killed the trees and the vegetation. The oil company canal, they let all the salt in there and killed the vegetation. And then it just ate away. We used to go out to the Violet Canal fishing. That was all wooded back before the ship channel. We would go out the Violet Canal to Lake Borgne. I remember when they came and cut an oil company canal. We used to go into the woods in this new canal and fish freshwater there. Or we could go out the Violet Canal to Lake Borgne and catch saltwater. There's no woods there any more. The new highway from Judge Perez Drive that crosses the Violet Canal—you can look out there and see the few dead trees. That was like a forest at one time.

Lysha (D), a fifty-one-year-old social worker: I mean we make these canals do things that we mess up. It's man who has done it. And it's destroying the land. We have progressed so much that we think we can do anything. But Mother Nature is going to turn around and show us we can't.

Many placed a large portion of the responsibility for land loss at the feet of the oil and gas industry. Equal to the application of responsibility, many thought the oil and gas industry owed retribution for their role, especially as the industry was perceived as having some knowledge of the damage it was causing. Christopher alludes to the commonality of this knowledge and the ripple effects begun by the oil and gas companies who cut canals, allowing saltwater intrusion to extinguish land and widen canals, bayous, and rivers. The transformation of land to water allows boats to speed through these waterways, thus increasing the wake from water vehicles and, in turn, further eroding the banks of the waterways.

While Christopher professes the normality of his expertise, Becky and William illustrated their insider knowledge by speaking about the loss of flora they have witnessed and which they believe the industry should repair. Lysha, on the other hand, reflecting the themes Gerry expressed in the last section, spoke of the consequences of human hubris and that "Mother Nature" would demand restitution for the debt she is owed. One wonders if Lysha believed Katrina and Rita were a partial repayment of that bill. Nonetheless, the perception by community members of undisputed fault was offset by a view of the industry as a local economic developer.

> Roger (D), a forty-seven-year-old commercial fisherman: They [prior generations] lived off the land. The whole family did. So I mean we are pretty much in tune with what it was and what it is now. They've seen it all. But the older people claim that the oil companies came in here even before the canals were dug. When they first started there was so much oil spillage. They claim that's why it's eroding now. Because it killed all the grass and then the bank just started eroding and it started from there. Back in those days, an oil spill, it wasn't nothing. They didn't have a DEQ [Department of Environmental Quality] back then and nobody was policing that. It was a money-making deal. They weren't worried about the environment like they are right now.

> Theodore (T), a forty-seven-year-old manager in an oil-related industry: Could I be bitter at the oil company for coming here and wasting the land so viciously? No. They gave me a good living. The house you are in today was bought with oil money. The vehicle I ride in was bought with oil money. So, we are a product of our own demise so to speak. The opportunity was there; we worked it. We didn't see harm in it at the time. But now these many years later, hindsight shows we may have damaged the land where it is irreversible. We pretty much have to say we are going to draw a line in the sand and say, "This is where it stops." Will it stop there? Not with the earth and the damage.

The oil and gas industry occupied a curious area in people's narratives. Like the others of this set, Roger's passage depicted an industry allowed to run amuck due to lack of regulation and the pursuit of profit. Roger placed this portrayal within the context of those who were firmly entrenched in place. Further delineating insiders from outsiders, he stated that it was his family and a community of elders who held a special relationship with their environment; they were "in tune with what it was and what it is now." Theodore also recognized the damage that the industry inflicted upon his environment, while he acknowledged the economic benefits it brought to him and his family. He conceded the benefits of the industry, but his use of the word "viciously" attached a brutal element to the meaning he intended and further implied a sense of guilt over his part in the damage.

This community accountability for oil and gas damage was not universal among residents, but it was not uncommon either. They represented the industry as an actor who might have known the harm it was causing while they, the residents, were lured by economic gain. The temporary gain residents incurred may have come at the expense of what they believe is their special relationship with place. While many projected the responsibility for rectifying the situation onto the industry itself, most, as Theodore and Lysha implied, believed this damage would be hard to remedy. In other words, many blamed the oil industry, but some also acknowledged their role in the degradation. In recognizing their function in land loss, community members realized, at least in small part, the damage that that may have done to their own relationship with place.

In contrast, the Mississippi River Gulf Outlet (MRGO), which runs through the whole of St. Bernard Parish, was not welcomed by residents. The outlet, completed in 1965, was mired in controversy from its onset, and became a symbol of the parish's land loss.

> Duke (SB), a fifty-three-year-old small businessperson: You've got a whole lot less wetlands because of saltwater intrusion due to horrific federal projects like the Mississippi River Gulf Outlet. You've got the Houma Navigational Canal in Terrebonne. You got all the pipelines dug by the oil companies. So you got land naturally subsiding. You got salt water coming in. The salt water kills a lot of the vegetation. The vegetation that required more freshwater, and things convert from marshland to open water. I forget the formula but it's like for every mile of marsh, absorbs a half of foot of tidal surge from a storm. The islands out there, the Chandeleur and Bretton Islands, slow down a storm surge.

> Christian (SB), a forty-two-year-old commercial fisherman: Going back to erosion—man-made—we spoke about this earlier, about the Mississippi River Gulf Outlet. Here you have a man-made monster that they dug this channel in 1962, I think it was. . . . The ship channel on an average is thirty-eight feet deep. . . . These ships come through this channel drawing thirty-eight foot of water in thirty-eight foot [water depth]. So they are touching bottom. They are dragging the bottom. When

that swell sucks up, I'm not exaggerating. I have been there. I found a ship coming down a channel, and if he don't have an eight-foot suction along those banks, he has none, I seen it. And there are sections in that channel where they rocked off after the protection levee.... But it keeps sucking from underneath and the rocks sink down. But for what I see on the top of the land, wherever they didn't rock, it's ate up. If it ate up in ten years three hundred feet, what is it going to do in twenty [years]? And they keep digging and digging.... They keep sucking that water and the land just keeps sinking.

Duke and Christian expressed their own specialized knowledge about the shipping channel known as MRGO. Duke placed MRGO within the larger equation of Louisiana's land loss. He connected the "horrific federal" channel to the degradation by oil companies and the deterioration of land that leaves his community increasingly susceptible to storms. Christian's focus, on the other hand, was more particular. Christian used his insider knowledge to give a detailed explanation of how the channel caused damage through drag by large ocean-going vessels and continuous dredging that was done to keep the channel operational. As was noted in chapter three, in spite of the continuous dredging, MRGO was rarely fully operational throughout its lifespan.

Hank (SB), a sixty-four-year-old government employee: What was back then is no resemblance of what's here today. And the MRGO has been the major factor. Subsidence is a natural thing that happens. But in this parish the Mississippi River Gulf Outlet is the main factor. If you want to talk about environmental terrorism, that's it. You are talking about degradation, death of a community. There's nowhere in the history of any community that the environmental damage that was caused because of the Mississippi River Gulf Outlet has ever taken place. This is it.... All there is is dead cypress trees. They stand tall like tombstones in a graveyard. That reminds me every day of what things used to be and what people will do for money. Greed.

MRGO was the big one for the people of St. Bernard Parish. Over the years it came to be seen by residents as a project forced on them by outsiders who promised it would bring economic prosperity to their communities but, in reality, only brought degradation. Additionally, actions to keep the channel open were viewed as counterproductive to projects meant to mitigate land loss. In Christian's passage, he noted that he saw how the dredging of MRGO dislodged rocks that were supposed to impede erosion. In light of the fact that citizens felt that their knowledge was largely dismissed, they regarded these types of actions as fantastically perplexing, resulting in descriptions that used words like "terrorism" and "horrific." The "man-made monster" that Christian spoke of was brought to a peak of symbolism by Hank, who viewed the channel as an act of "environmental terrorism" bringing continued degradation and a slow death to the communities and the land where the "dead cypress trees . . . stand like tall tombstones in a graveyard."

None of the interviewees from any parish viewed the Mississippi River Gulf Outlet positively. In fact, nearly all St. Bernard residents mentioned the channel, and all of those perceived it in a negative light. The channel occupies a large space in their symbolic landscapes of coastal land loss. Even now that it is closed, it is a construction that many come into contact with daily and is the ire not only of those in St. Bernard but those in the surrounding parishes as well. Ironically, if not for MRGO's aid in channeling Hurricane Katrina's storm surge water, the resultant levee breaks and flooding, and then the subsequent regional outcry to close the channel, it might remain open today.

Those not in St. Bernard tended to view the outlet as another instance of human degradation in a long line of negative impacts upon their environment. And while the harmful actions by humans were reflected in very localized experiences, residents also viewed humans' impact on a larger level.

> Bubba (D), a twenty-seven-year-old recent college graduate: And I think it has everything to do with the control of the water flow over the years that has affected—we had a really bad problem. I think the marshes are meant to be—Mother Nature intended them to be brackish. And she controls them by allowing a certain amount of salt water to infiltrate the marshes, and yes, it does kill certain vegetation. It allows certain vegetation to grow. In our marsh it used to be very brackish in nature, and so the marshes used to be strong and vibrant.

Next, Anastasia (GI), a twenty-seven-year-old education professional, was asked to further discuss her perception of the mismanagement of the Mississippi River.

> I should be looking more at the science of it right now to know exactly what's going on. But my perception has always been that when the Mississippi [River] has been managed to benefit the people of New Orleans and the oil companies, they cut canals through marsh without any concern about what kind of impact it's going

to have. And eventually what's going to happen is the marsh is dying, and I remember, just in my lifetime, this beautiful green expanse between ground and the road being full of water. It used to be an actual living system.

In addition to the meaning residents attach to their direct experience of the event, the experience of coastal land loss is given larger, more abstract meaning. Other than commercial fishers, most residents do not see the direct causes of land loss such as canals cut by oil companies or land eradication by storms. They mostly see the indirect effects such as dead cypress trees and wild birds in urban spaces. Yet, these localized experiences were given meaning that was extrapolated to a larger disastrous event they viewed taking shape.

The way they described the disaster and the larger context they placed it in were also products of their damaged attachment made all the more substantial by its human dimension. Residents blamed powerful outsiders but they also accepted responsibility for their role. Many believed that, in pursuit of better economic circumstances, they were willing participants in the cause of coastal land loss. However, in general, and possibly a result of the heightened anxiety many experienced, they placed blame on government and coercive economic actors. Their raised angst, along with continuing land loss and the tension between their own culpability and their subjectivity to more powerful forces, led to another unwelcome insight: certain doubt about the future.

UNCERTAIN PLACE, UNCERTAIN SELF

In large part, loss of any sense of stability defined how people experienced the phenomenon of coastal land loss. Add to this instability the belief that the disaster is caused by outsiders and that ambiguous and piecemeal solutions are implemented by disrespectful others, and the result was residents who were uncertain about place and identity. Of course, there were glimpses, sometimes substantial, of this doubt in the previous themes. However, it is in this theme about *uncertainty* where the narrators' intentions of ambiguity and insecurity are given full light.

> Susan (GI), a thirty-year-old graduate student: I think I have a much better ability to ignore things. But when I came back, when it really hit me is when I came back from college. I walked on the beach and realized there is no beach. I have to walk in the water. It's scary. It kind of affects your sense of well-being; it tests your sense of

well-being. All your life you are expecting the beach to walk on and run on, and you step off of the levee. It almost feels like you are stepping off of a cliff, a very short cliff when you grew up seeing sand you could walk on. I've had nightmares about just dropping off. There's nowhere to go, just down.

Susan's symbolic landscape illustrated the threat to identity that land loss had for her. She noted the effect it had on her "sense of well-being" and then repeated the sentiment, adding weight to its significance. Her sense of self, given solidity through place, was thrown into question due to the disaster. In other words, she knows who she is because of where she is from. When that place begins to disappear, the sense of identity that was once secure is now uncertain, so much so that the threat to identity invaded her dreams.

Throughout these pages residents have demonstrated the anxiety that results from their experience of losing place. So, intuitively, it might seem that plans to alleviate the disaster would bring some comfort. However, as their narratives have revealed, their worries were not assuaged by plans to build coastal land. The danger to identity that land loss poses was further demonstrated in narratives about what may be done to lessen the deterioration.

> Carrie (GI), a forty-six-year-old government employee: I really don't know what can be done because I know they put the rocks and stuff and that might help it here, but right up the road, when you're leaving the island, just take a look on the right-hand side and it's like every time I go up the road to go do some shopping and everything, I'm amazed at the amount of water that's there and there's no more land. I mean, you don't realize, but just the last several years—I know it took years, but now when the tide is high, I'm like, it's really scary and when you get off the bridge, there was a lot of land and now when you get off that Leeville Bridge, there's water. They have some people that's come down and they haven't been down here for years and years and they got scared and turned around at the Leeville Bridge and said they couldn't believe there wasn't no land left. I really don't know what—they talk about different things, different organizations or different groups and committees and stuff and getting—I don't know, is it the silt from the Mississippi River and let it build in one area and so forth? I don't know if it's an ecology thing or whatever, but I don't know long term or whatever, because this is happening quick. I mean it's gone real quick. They say they're losing. I've forgot exactly how much land per year and it's—you could see it. I mean you could see it. Quick. It's going quick.

In the *restoration* theme community members expressed what they thought about solutions, which oftentimes translated into frustration and sometimes anger. But here they expressed their uncertainty. This uncertainty left them insecure, which, in turn, increased their sensitivity. And it appears likely that this sensitivity heightened their frustration over what they believed was a lack of action. While Carrie seemed frustrated, bewilderment was the point she wanted to get across. Confusion over what could or would be done was common and echoed often. Another example, Tina (T), a twenty-nine-year-old childcare employee, said, "There will be nothing if something is not done. And I don't even know where people would start to try to get something done."

For many, the uncertainty that developed from the disaster, as Carrie said, was "scary." It may have been that the incessant nature of the event caused residents to attach foreboding and confusing meanings to the phenomenon. Some, like Tina, found the situation somewhat overwhelming. Some had a "bad feeling" that water would overcome place, and some, like Art (P), a fifty-two-year-old government employee, expressed the shock of lay people in metropolitan New Orleans after they learned the severity of the problem. Consequently, place was viewed tenuously upon realization of the scale of land loss. In Carrie's passage above, she employed a somewhat similar anecdote, explaining that residents returning to her community of Grand Isle after a period of separation were frightened to the point of not even making it over the "Leeville Bridge" before departing again.

Although community members were expressing their uncertainty, an equal intention remained that of demonstrating the gravity of land loss. When they told stories of city dwellers or returning residents who were "scared" due to the amount of loss, it was meant to give credence to the severity of the event. In effect, people were saying, "So, you can understand my fearful uncertainty."

Nonetheless, community members' doubt was not directly due to land loss. More so, their unease was induced by experiences that were once familiar but now rendered strange by land diminution. The following seven passages illustrate the experiences that produced so much uncertainty. Place was no longer familiar. The locales that people identified with in the past, the places that carried the elements of a constructed identity, had disappeared or were disappearing.

> Theodore (T), a forty-seven-year-old manager in an oil-related industry: When you go out there and you running your boat and you remember seeing a piece of land one year and you go back and you say, "What happened here?" The landmarks that you used to use to navigate by are no longer there.
>
> Jackie (LC), a fifty-one-year-old fireman: I used to fish down in Pointe a la Hache, a lot, too, as a kid. And I remember just a few years ago, I went down to Pointe a la Hache fishing, and it was like I was in a whole, totally different [place], like I had never been down there before.

Many residents communicated an unfamiliarity of what was once known territory. There was a strangeness to place that made them question the future while still reaching to the past.

> Lester (SB), a forty-nine-year-old commercial fisherman: If the old people could come back now and you put them in the boat, they wouldn't know where they at. That's how much the marsh changed. There's places in the last couple of years I've got lost. It used to be land and bayous and now it's just open water.

Lester illustrated not only an identification with the land, but also that of an identity connected to past generations through place; a connection that he and his neighbors believed was in jeopardy. Below, Chuck and Tina echoed Lester's thoughts by relating the fear they and their elders held.

> Chuck (T), a thirty-six-year-old commercial fisherman and oil field employee: My dad is sixty-seven, sixty-eight years old, and every time we go to the camp, he won't get behind the wheel [of the boat] unless he has to. And he's been born and raised out there, practically his [whole] life.... Talking about having twenty feet of land right here. You put the PVC pipe on the end of the point [of land]. Come back a year later and that twenty feet is just about gone. It's weird. It's no way of stopping it that I can see.
>
> Tina (T), a twenty-nine-year-old childcare employee: It seems like everything is opening. And even like going out there [in the bayous] with my grandfather and him getting turned around and he didn't know. He grew up there as a child, and now he is in his late seventies. And when he got turned around, that's what scared me. I didn't know. That's not a route that I'm used to taking. But I knew that there

was oyster reefs around there. . . . But we need some help from somebody. I don't know where you would even start.

Statements like these held an almost desperate air. Those from whom they had gained knowledge and who taught them about the land and waterways now felt insecure and even frightened. Also, the sense of security that family elders imbued with their expertise was lost. Next, Rachelle (T), a sixty-one-year-old chef, inverted this theme of generational knowing.

I went out in the beginning of June. I went out fishing with my son. Now I hadn't been on the water in about five years. And I used to get in the boat and I'd just take off and go fishing. And he says, "Momma, tell me a few good spots." I said, "I'll take you there." I get there and I don't know where the spot is. It has changed that much in five years. And I said, "I can bring you close to it. Like this one place right on the other side right here." I mean this was always good for a few specs [speckled trout] anyway. But it's eaten away so much, I did not know where the reef was. It's changed. It has totally changed. It's sinking. It's washing away.

Her inability to give her son, the next generation, her local knowledge deprived him of the same connection with place that she carried, as well as knowing something more about who his mother is. These statements were given weight by her closing sentiments: "It's sinking. It's washing away."

Relating stories in this way was purposeful on the part of narrators. They wanted us to understand that the place they called home was most uncertain. For many of us home is a place we identify as a symbol of refuge; however, residents believed this sense of stability had been taken away from them. At the same time, speaking about severed generational connections and using words like "scared" were ways of expressing a deeper threat to identity, a threat that was occurring due to the disappearance of a place to which they felt intensely connected.

An interesting comparison with the statements of the last several residents is the next passage from Sven,[10] a thirty-three-year-old commercial fisherman and a resident of Delcambre, the community which lies farther west in the center of coastal Louisiana.

Because I go the same route and I have poles marked in the same place, you can actually see it. You got to go out every year and we mark it. Where's it going to be?

> You can see it on low tide and north wind. When the water gets really low, you can see where it's at. But on a high tide you don't know. You've got to go out and figure out where it's at this year.

Sven's comments lacked the gravity that the others gave the event. His experience of unfamiliarity was minimal compared with other Louisiana residents who lived farther to the southeast. Reflecting Sven's tone, the people of Delcambre spoke about land loss less than members of the other communities. Citizens from the community of Delcambre also spoke less about a connection between environment and self. Perhaps this was because they were less affected by the disaster of land loss than the other communities prior to the storms of the last several years.[11] Delcambre's divergence from the other communities also may have been due to the overall cultural distancing of humans from their natural environments in conjunction with a decades-long economic downturn in that community and, at the time of these interviews, the lack of an event to cause acknowledgment of place attachment.[12] Still, the rapid onset of events over the past few years, increased land loss, and the near assurance that these events will be more likely for the immediate future make it likely that the sentiments of Susan and Liane below now resonate with the residents of Delcambre. In her passage, Susan (GI), a thirty-year-old graduate student, was speaking about her father and residents like him. Going to school out of state, Susan had not been a resident of Grand Isle since high school.

> The feeling that everything is washing away and there's nothing to look forward to. There's no future. There seems to be a lot of nostalgia and a sense of sinking into the ocean personally. Just watching everything disappear and you can't do anything about it.

> Liane (T), a thirty-eight-year-old community organizer: I don't know what to do about it. I don't know what anybody can do about it right now. I don't even know if it can be saved. To me this is an emergency. Just like a patient, the land is an emergency to people. That's their livelihood.

If the passages of some of the other residents like Tina, Chuck, and Rachelle only bordered on desperation, there is little question about the desperate element in Susan's and Liane's narrations. It was more than not knowing what the future would hold. And it was not only the land "washing away" but the people who, as these and other passages suggest, belonged to the land and who were like an "emergency patient" that was in grave danger. As Susan indicated, coastal land loss reflects an end to identity, a metaphorical death.

Brown and Perkins (1992) point out that it is only in retrospect, after loss, that people acknowledge the depth of their attachment to place. Citizens of coastal Louisiana were and remain in the midst of this loss. They "see it" occurring slowly every day, yet at the same time it is "quick" and deceitful. Because of the continuous nature of the disaster, these residents appeared to be perpetually aware of their attachment. Carmen's thoughts from the end of chapter four exemplify Brown's and Perkins's assertion as it relates to Louisiana's phenomenon. Her statements are not postevent but during the extended event.

> Carmen (T): Never take for granted that the land that you are on will always be there. Never take it for granted. It disappears in an instant. Never take for granted that you can put something in one spot and when you come back, a couple of years, it will still be there.

Just as the loss of land continues, Carmen's statements remain a warning for others who are as attached as she and her neighbors. She expressed the simultaneously slow yet instantaneous nature of the loss. What may take "a couple of years" to disappear might feel like loss "in an instant."

Residents experienced not just uncertainty of place but an uncertainty of identity as well. They attached meanings of helplessness, anxiety, despair, and strangeness to the event. They did not limit these meanings to the demise of the land; they also adopted these characteristics into their self-definitions. Greider and Garkovich (1994) state that when a place changes, a renegotiation of self and identity occurs in relation to the changes transpiring in that place. The damaging meanings that were given to coastal land loss were also conferred onto identity. These meanings are continually adjusted and transformed as the event alters and transforms. As a result, the continuation of the disaster generates a slowly emerging new identity. As these narratives conveyed an identity that was struggling to find some stable ground, the narrators found no assurances from government, the institution that was meant to provide that very function.

POLITICIZED NATURE

Interestingly, uncertainty among residents seemed to solidify as the political elements of coastal land loss were played out. In other words, people became more certain of uncertainty due to their distrust of government and politicians. Residents have watched this ongoing disaster for over three generations. Yet it is only over the past fifteen to twenty years that the event has slowly gained widespread political attention. Most community members were highly skeptical and cynical of this process, yet they held out some hope.

> Jenny (GI), a fifty-four-year-old who works in public relations: I hope that the government continues to do what it does so well right now—lobbying Congress and the state for the help to preserve us physically, as well as our history—because when Grand Isle is gone, New Orleans will be gone.

Again, there was a call for a political response framed in warnings about the dire consequences that could result from land loss.[13] Even so, Jenny's warning was contextualized within her mostly positive sentiments. However, the assured and hopeful meanings she gave to the political component of land loss were rare among residents. While many more expressed a moderate degree of hope, it was usually offset by negative narrations of how politics is part of this ongoing disaster.

On the other hand, there were those, although relatively few, like Walter (T), a fifty-one-year-old facility superintendent, who took a more realist stance toward the political climate.

> What I see and the way things are going for funding to rebuild those areas, I think they are beyond repair. If they can slow it down, they'll be ahead of the game. The money it would cost to rebuild would be too much for the rest of the country to go along with. And we can't do it ourselves.

The symbolic landscape that Walter constructed posited that he and the people of his region cannot solve a problem that is "beyond repair" without large-scale political will. Politically, Walter viewed Louisiana's coastal damage as a regional or state problem. However, he was not fatalistic; he employed a sense of realistic hope that the problem might be slowed. In the next two passages, residents expressed a more dismayed sense of political obstacles.

> Rachelle (T), a sixty-one-year-old chef: Give politicians something to work with and you know what happens. Because, I mean all these years, they have been talking about it, and they really have not done anything. Nothing has been done. They'll say they are trying something but it doesn't work. As long as they are trying, I'm happy. But then they hold another survey and let it drag on again. So by the time they could do something, things have changed. So they need that other survey to figure out what to do again. It just keeps going.

> Liane (T), a thirty-eight-year-old community organizer: I just cannot believe the government has watched this happen to our people. I just can't. It's not just our area, just a coastline.... And how our state government can sit there and our senators, Congress can just watch it go. I just don't understand it.

These community members were more skeptical of the political process than Walter and Jenny. Of these two passages, Rachelle's comments were more common. Many saw politicians using the issue of coastal land loss for political clout, but where no substantial work had been implemented. This belief was projected onto the agencies that are charged with fixing the problem. Others, like Liane, expressed bewilderment about government's involvement or lack thereof. Her words illustrate how political inaction fed and heightened citizens' anxiety and confusion about the future. Liane gave weight to her puzzlement about government by attaching herself and her neighbors to the land and regional coastline. In this way, her intention was to imply that the people and the restoration needs of the region were neglected by those who had a hand in its decimation.

While some thought government was neglectful, many residents also aired frustration over the bureaucracy that arose around coastal land loss. Bettie (LC), a fifty-six-year-old homemaker and local historian, speaks about the application process for land mitigation on private property.

> But in order to put something down whether it's sheet pilings or it's concrete or it's trees you have to go through so much bureaucratic red tape that it stinks.... And this wasn't like I could call up on the phone and say, "Hey, I wanted my forms sent

out. Could you get them to me 'cause you know we're watching the water slowly come up to the front of the house." You're talking about six months, eight months; you're into hurricane season. You're into another foot or two [of land] gone away. So know that if you put back more [sheet pilings, concrete, or trees for mitigation] than what they told you you could put back then they come out and go, "Well, you're gonna have to remove this because we only gave you permission to put four feet and you've got eight feet." "Well, asshole, it took you eight months to tell me, in the meantime we had two hurricanes." You know? "A lot of boat traffic so we lost another two feet while you were screwing off someplace." It's a bureaucracy. That's what it is. The shit you have to go through is phenomenal.

In Bettie's passage, we see again the insider/outsider dichotomy that was prevalent in some of the other themes. Bettie's self-definition was that of an attached insider who wished to act as a steward of her piece of coastal landscape. She and her fellow residents were trying to sustain their homes and land while different government agencies not only did not help but obstructed their attempts. Interestingly, there was also the implication of a "we," in which she and her similarly attached neighbors possessed some solidarity due to oppositional bureaucratic institutions. In this way, the phenomenon was not just a "natural" process of wetland erosion. That is, the definition of coastal land loss also consisted of outside political agencies that added to residents' problems.

> Albert (GI), a thirty-four-year-old manager: Because I had one fella [from the U.S. Army Corps of Engineers] tell me that they ought to take the money for erosion and put that to transplant the people, to move them. Instead, if your house goes, well, don't rebuild Grand Isle, just take them and move somewhere else. So I told the fella, I said, "Well, okay, that sounds—" I said, "I'll be there at your house tomorrow with a U-Haul." "Oh, for what?" "Well, I'm moving you and where I move you, you have no—" "Oh, there ain't nothing wrong with my house!" I said, "Oh, don't get upset. You just said you want to uproot me. So, it's in good faith that I go move you wherever I want to move you, you know." Not just get you out of the place that you call home, you know. . . . Well, whatever I mitigate, what I damage, I have to mitigate. Now, all of a sudden they've come up with a price tag on how much marsh, how much it's worth. And yet, when you turn around and say, "Okay," and tell them to play by their rules that they've made, they don't want to. I said, "Well, all right, why don't you do that?" "Well, uh, we don't have the funding." "What do you mean

you don't have the funding? So where's my tax money going?" You know, we've fallen into that and seen the bureaucratic stuff that's just run around in a circle more or less because nobody wants to agree on anything, you know....[14] We're still going for it. Unfortunately it's made me bitter about Wildlife and Fisheries [department], National Marines and all this, organizations that nobody has, knows anything about this area, but yet they're going to tell you what to do, what you can't do to the area. But they don't know nothing about it. And it's, it's made me very bitter about it. That's like [U.S. Army] Corps of Engineers, you know it's like, they have stuff. They keep griping and he's [U.S. Army Corps of Engineers representative] telling me that if we dig into the island it's going to mess it up and this and that. And I told him, "What about the east end of the island? It's about to be cut through by the state park, what are you all going to do about that?" The man flat out told me, "I don't want to talk about it." I said, "Well, why not? This is in relation to Grand Isle, and you can't tell me what y'all are gonna do about it?" I say, "I just want people to cut—don't worry about people cutting through the island." In fact, that was the man that told me that we ought to just uproot everybody on the island instead of worrying about coastal erosion.

Albert's extensive retelling reflected his unique position as a developer and lifelong coastal Louisiana citizen. The meanings he gave to the loss of coastal lands were filled with conflict and distrust. He illustrated the frustration he experienced with the government agencies, cast as outsiders, who, he believed, were supposed to not only help restore the land, but assist community members, as well. Perhaps he held this belief because he thought himself to be as integral to the coast as the land that was supposed to be saved. However, these concerns are dynamic and don't occur in a vacuum.

In Albert's case, he was a partner in some commercial development in the area. In accordance with state and federal regulations, any new development in coastal wetlands must be mitigated. The permitting process can be long, complicated, and may involve multiple agencies. Once development is approved, the builder or owner agrees to replace wetlands on the same property or in another area. Part of the reason for Albert's frustration might have been the perception that the permitting process didn't occur the same way for everyone. Many believed that the larger the landowner or company the smoother the bureaucratic road, with little to no mitigation responsibilities. Environmental and resource agencies that stood in opposition to individuals and communities were an obvious concern for Albert, but most residents maintained this view. Considering the strong identification these people have with the land, it is no wonder that they assumed that land restoration and community assistance should go hand in hand.

> Bettie (LC), a fifty-six-year-old homemaker and local historian: I think if the corps would just—the [U.S. Army] Corps of Engineers are a bunch of assholes. They are total assholes. They are a bunch of educated asses that don't see what the factor is. It's like you sitting behind your desk and saying, "Well, I think the packaging of bread should be changed." Fifty thousand people have bought bread in that packaging for fifty years, but because you're sitting behind a desk, all of a sudden you decide it should be changed, it could be changed. They don't look at an overall picture. My husband deals with the corps all the time because he's a machinist, because of the locks they work together a lot.

> Paul (P), a forty-two-year-old commercial fisherman: And then when you see the hypocrisy of the state regulators and their policies, you know for three generations, we've been fighting the coastal restoration effort and the political in our life. My uncle's got relatives from 1930s and 40s that he was talking about, "Y'all better do something because it's going to progress." And I was telling him, "We are the outlaws and they are the experts." When it was [happening], we had been trying to bring this attention to all the devastation. The oil company was going out there and natural processes that were going over there. Because if the lands, if it dies, a part of us dies. They don't seem to make that connection inside.

Coastal land loss was politically charged. Bettie's and Paul's words revealed the assault to identity they experienced from agencies like the U.S. Army Corps of Engineers. Citizens viewed these agencies as politically motivated and mobilized. This politicization of land loss worked against residents' sense of well-being. While Bettie used strong language and an analogy exemplifying the dismissal of community involvement to convey her experience, Paul thought that through politicization, community members had been cast as the "outlaws," into criminals on the land. He contextualized this statement within his strong ties to place. At the beginning of the passage, he mentioned his family's generations-long battle against land loss, and he ended his narration by reiterating his community's strong connection to the land. He pointedly established how much this coastal place is a part of who he and his fellow residents are. Establishing this strong bond and then contrasting it with the inability of bureaucrats/outsiders to grasp this, Paul gave significance to the battering of identity he experienced from government agencies.

The political component of coastal land loss was echoed by many where government was viewed as a collection of powerful and manipulative colonizing outsiders. Taken together with the other components of coastal depletion, this politicization alienated residents and compounded their frustration and anxiety. In the next pair of passages, St. Bernard residents spoke about the U.S. Army Corps of Engineers' Caernervon Freshwater Diversion Project, which diverts freshwater from the Mississippi River into the marshes of St. Bernard Parish. The project is meant to deposit sediment from the river and build land.

> Phyllis (SB), a forty-seven-year-old office administrator: I wanted them to explain to me how could people like us benefit by it like they were telling us it was going to do. Everything is a big lie. And people down here, they're not that stupid. You might get a handful of people that say the freshwater [project] was the greatest thing that ever happened. To me I can't see where.... They don't care. If you are Indian and they want your land, they go to the reservation. If they want your house, you need to sell it at the price they want to give you. The money's in the bank. The interstate is coming through. That's the way the government does it. Look what happened to the Indians. So what made them think that by saying fishermen—this was a cultural thing, and they had to help save this because of the cultural purpose. They tried that route. It didn't work.

> Tyronne (SB), a forty-year-old commercial fisherman: I think they just eating up the land, washing it away. My opinion, I think it's all about money. They getting so much money to run that thing every year. So they showing whoever they got to show these pictures of the grass and tell them it's building land, and they collecting money for it to run it. And that's what it's all about. They're really not helping the land. Somebody is making plenty money off of it. They had people from down here since they put the siphon [in], brought them guys that's controlling that siphon for a ride in the boat and showed them pictures of canals and how it was so many years ago and how much bigger it is now since the siphon and it still don't do no good. They just showed us this grass and made like it's filling up all kind of land. You jump in it, you go over our head.

Deceit was the meaning here. Phyllis, married to a commercial fisherman, appropriated the historical fraud and eradication of Native Americans by the U.S. government to convey the extent to which she believed the current government was imposing its will on her community. Meanwhile, Tyronne suggested that competitive marketplace processes and deception propelled the bureaucratic business of coastal restoration. Again, the agencies and political processes that are supposed to be on the side of residents symbolized a disaster in and of itself. The next resident reiterated the ideas above.

> JJ (T), a fortyish marina owner: I think the bureaucratic world is going to continue to procrastinate, continue to talk, continue to use it as a platform to get elected. They will continue to use it as propaganda, as gossip. And from there they will continue to study. Because once they solve the problem, the money source may dry up. So they will continue to study, they will continue to procrastinate, they will continue to use scientific data, and money will float. It's the float of all the channels of all these people who are professing to solve it but are not really concerned about it. And money is going to continue to—until the money is put down in viable projects, in viable solutions, until the solutions are being solved. No matter how much money, no matter how much talking, no matter how much hoopla you get. It's good pretending, it's good news, it's good to write about. It's making everybody millions of dollars, but it's not taking care of the problem.

JJ attached meanings of pervasive "propaganda" and bureaucratic profiteering to the political atmosphere of land loss. It was not only common for residents to frame the government and politics negatively, but it was also common for environmental government agencies to be characterized in the same way, as propaganda machines for the benefit of politicians and themselves, not citizens. In fact, community members partnered those agencies with the more traditional politicians, casting them all as being in cahoots with one another and working against coastal restoration and community preservation.

Although not as prominent as the first couple of themes of this chapter, residents' symbolic landscapes held a strong political component that when expressed was profound. Not only did experience with the daily disaster of coastal depletion affect communities' personal relationships with the land and definitions of self, but those encounters were also colored with political overtones. So, when they noticed another piece of land that had disappeared they also felt the brunt of political ineptitude. Yet, even though some were hopeful about the political process, they and most others saw discrepancies that led them to be highly skeptical about a process that may be their only hope for land and community sustainability. This section closes with a short passage that seemed to sum up residents' general views of what they believed was the political inaction of coastal restoration.

> Morris (T), a twenty-six-year-old recent college graduate: I don't know [what will happen]. Some more studies. The rocks [rock jetties] seem to be working well, but nobody wants to put rocks. They want to study it. It's too much politics and too little action.

CONCLUSION: EXPERIENCING FRAGILITY

Simply put, experiencing coastal land loss caused residents to express a sense of fragility. People attached meanings of vulnerability to the phenomenon, and all of the themes, including the effects on attachment to place covered in chapter four, communicated a sense of fragility about place and identity. Consequently, outsiders and the process of restoring the coast were also experienced anxiously.

Including chapter four, six themes emerged out of interviewees' discussions of coastal land loss. In a general way, these themes encompass the intended meaning that residents gave to their experience of land loss. The meanings reflected community members' self-definitions, sometimes intimately, sometimes in a detached way, but meanings were always attributed to land depletion through the self as it related to their communities and the coast. These meanings constituted their symbolic landscapes (Greider and Garkovich 1994).

The narratives of these residents make it obvious that the loss of coastal land is a personal issue that affects their sense of who they are, their identity. In addition to the effect on identity, and perhaps, in part, because of it, land loss took on many layers of meaning. As a result, sometimes community members were unclear or wavered in their thoughts. In other words, the event was not just an environmental change occurring where they lived. Residents communicated attachment and fragility by connecting childhood memories and familial connections to land that no longer exists, having knowledge acquired over a lifetime challenged by incongruent events, connecting historical identity to eventual displacement, and sudden awareness of "the problem," to name a few.

Indicating their direct experience, people repeatedly reported that they *saw* the loss of land. Seeing this loss symbolized a slow eradication of a way of life that they viewed as dependent on place. It was not only their physical proximity, but residents' cognitive closeness with the coast that gave them a sense of bearing eyewitness to a process that others only knew in an abstract fashion. Paul (P) sums up the closeness they felt not only to place but the loss of it when he said, "Because if the lands, if it dies, a part of us dies." Residents conflated themselves and place. The line between themselves and the land, if they saw one, was blurry and elusive at best.

In addition to the personal elements that community members acquired from the forfeiture of land, the phenomenon took on a political affect as residents directly or indirectly experienced a diverse array of governmental organizations that, for the most part, left them frustrated and resentful. Land loss was further politicized as they viewed governmental agencies using the event to employ and build further political capital.

Experiencing this disaster meant seeing little done to solve the problem, while being disregarded, undervalued, pushed aside, and taken advantage of. Perceiving these elements from outsiders produced a robust distrust and skepticism among residents who were already fragile due to their significant intimacy with place. Individual and community approaches to preservation and restoration served to reestablish some sense of control, hence, reifying connections between the land, home, and identity. However, when resorting to their own methods, residents met with opposition from an array of seemingly contradictory organizations.

While it is important whether residents' actions were actually restorative or, in the long run, detrimental to the land and their cause is not the point here. The point is that many citizens believed that not only were they not being helped or listened to, but in many ways government was making matters worse. Consequently, when agencies and residents came into contact, at least as reported by these citizens, conflict was likely to arise. Conflict occurred as agencies proposed restoration projects and then increased as residents attempted to implement their own mitigation measures. Nonetheless, residents believed in their own agency as exemplified by Jackie (LC) when he said, "If it wasn't for the people out here who does fill in and put mud in and put 'wash out,' if they wouldn't do that to the land out here, this island wouldn't be here anymore."

Coastal Louisianians faced challenges on two fronts—from the disaster itself and from the restoration process. Based on the attachment of residents to place, the alienation from and perceived inaction of the restoration process, as well as generations of community land loss mitigation, it seems reasonable that residents would take matters into their own hands.[15] Community members were aware of land loss long before any official restoration process, and they took action such as they could. But the bureaucracy of preservation and restoration, as well as the scale of the disaster, rendered individual efforts almost meaningless, adding to their sense of fragility. Nevertheless, many stated that they engaged where they could, on their property. In short and generally speaking, the restoration process symbolized inaction, disrespect, conflict, condescension, alienation, and distrust of government and its related agencies.

Conflict with the governmental element of coastal land loss was viewed as a kind of social damage that aggravated the ecological and personal damage of the phenomenon. Most residents held a generations-long connection to the region. Their identities are still bound to it. Even nonnatives expressed a strong connection and identification with place. Residents conveyed a post-disaster acknowledgment of their place attachment, a heightened awareness of their connection to the land and the region (Brown and Perkins 1992). However, due to the incessant, ongoing nature of this event, interviewees expressed this heightened awareness of attachment in the midst of coastal land loss. In fact, looking at the words of community members, one can see that it was their intention to communicate such a strong and nearly inseparable attachment.

As stated earlier, the research of Brown and Perkins (1992) finds that disasters force an acknowledgment of place attachment. In times of relative normalcy, people are often unaware of the level of their attachment. The authors state that for a period immediately after the disaster, a heightened awareness of attachment is experienced due to actual loss or the possibility of loss. Building on and extending this finding are Louisiana's coastal residents who, it appears, experience a more constant awareness of their attachment to place. In other words, they are always at a relatively heightened awareness of place attachment. It may be impossible to constantly be in such an aroused state—it most likely ebbs and flows—but because of the slow yet incessant nature of Louisiana's disappearing land and the different definitional elements that the phenomenon takes on, they are more often than not aware of their strong attachment.[16] Furthermore, as residents said themselves, the self *sees* the ongoing damage, thus reinforcing the notion of their more constant awareness of attachment. It was the constant awareness of attachment due to a slow disaster that produced the fragility people said they were experiencing. In fact, they experienced a more constant awareness of attachment due to the nature of the meanings they gave to place, and this awareness caused the fragility of place and identity that residents experienced.

From such a strong attachment and identification with place, damage to their environment becomes injury to the self. Harm to a healthy and nurturing ecosystem is harm to the self and therefore to the components that make up identity. Decimation of their environment, realized in such things as dead and dying trees, comes to symbolize the dire threat to identity. Residents faced the possible death of an integral component of identity. Thus, the potential peril identity faces causes the phenomenon to be experienced through anxiousness, desperation, and vulnerability.

A disappearing healthy environment, along with an antagonistic political process and suspicious business practices, conspired to yield meanings of great uncertainty about the future of this place to which community members are deeply connected. The anxiety that this uncertainty produced was often expressed as not knowing what could be done, or if anything would be done, in questions about whether or not place would exist in the near or distant future, and in repeated verbalizations of previously held but now lost geographical knowledge of place. As this anxiety was exacerbated by acknowledgment of their attachment to place, uncertainty became expressed more despondently. Death became a popular metaphor and calls for salvation took on an air of desperation. It was from here that identity appeared most fragile. Residents' calls for assistance expressed the anxiety they experienced as a result of their damaged environment and identity. The rhetorical call for action reflected the fragile nature of identity, and, while many times explicit, the call was also metaphorically implicit through such images as death and disappearance.

However, it was not only uncertainty about place, but the sense of who they are that was bound up with place and thrown into question. As their symbolic landscapes took on new meanings in accordance with the fluid and changing nature of the event, residents' self-definitions underwent constant renegotiation. Although to some extent all of our self-definitions are fluid and changing in relation to changes in our environments and to the extent that we are attached to them, the renegotiation of community members' self-definitions took place negatively within the context of a continuous disaster.

Perhaps being asked to construct a narrative about place forced acknowledgment of the depth of possible loss, thus causing people to broach the subject of land loss and to speak about it in such dire terms. In other words, through residents' own comments, we know that they were, more or less, always aware of land loss. Being asked to compose a narrative of place may have caused residents to talk about land loss because of the very threat it posed to that narrative. Residents' symbolic landscapes, the meanings they attached to the coast, consisted of a personal narrative that was framed by the natural and physical elements of place. Land loss was threatening the very existence of that narrative and forcing a renegotiation of self-definitions. In this way, the disappearance of land was not just another element of meaning added to their conceptions of the coast, but one that invaded every component of their conceptions. This invasion rendered identity fragile. As a result, to experience land loss means the possible loss of identity—past generations, future generations, and the general sense of who one is. This caused a heightened sense of anxiety expressed in calls for salvation both from within and without, explicit and implicit.

In sum, experiencing Louisiana's coastal land loss coalesces several elements producing a fragile sense of identity. Coastal residents are deeply attached to place, and this alone caused changes in how they viewed their communities and the region that had ramifications for identity. Add to this the damage they experienced on a daily basis along with the looming threat from hurricanes, the negligence of mostly powerful outsiders regarding this damage, and an alienating, distrustful, inactive, and politically oriented restoration process, and identity came to be defined with uncertainty. When residents looked at land loss, it was through these different yet symbiotic elements that the event was defined and experienced. Coastal land loss became a monolith that affected all parts of residents' symbolic landscapes.

So the question becomes, what can be done to lessen their anxiety and fragility?

6

SAVING PLACE
Residents and Their Environment

Why should alleviating Louisiana's coastal residents' anxiety and fragility be a concern or goal? This might sound like a rhetorical question except for the fact that many would proclaim that this anxiety and sense of fragility is an unfortunate outcome of any significant change. Their refrain would be "change is hard," and they would liken the stress to growing pains that will recede as people and society adjust and account for these changes. In her eloquent foreword, Sara Crosby referred to this in her fellow Grand Islanders' reluctance to ask for assistance due to the retort "Why don't you just move?" and all that that question implies about being a victim.

Nonetheless, as we seek our reflection in increasing democratic principles, a first answer to this question seems to be a certain inalienable right to live with a modicum of trust in the capacity of the immediate environment to support a certain quality of life. Second, and more practically, but related to this expectation, coastal Louisianians' anxiety and fragility leads to conflict with agencies charged with coastal restoration. Further, and an extension of this conflict, there is opposition to coastal restoration policies before they ever get to the implementation phase. The onset of Hurricanes Katrina and Rita is unlikely to change this conflictual relationship. It may even exacerbate it. If disasters cause a heightened awareness of place attachment and these residents were already in a relatively heightened state of awareness, then Katrina and Rita gave credence to their raised levels of place possessiveness. Place and identity are viewed as further threatened by storms such as these, whose damage, they perceive, in part, as a result of the ongoing disaster of slow-onset coastal land loss. Some blame for the storms' destruction has been laid at the feet of those agencies which coastal residents believe have taken little action to alleviate the loss of land. Thus, it is unlikely that they will blindly accept new or additional proposals from agencies they deem untrustworthy and threatening to their place attachment constructs.

By no means does this suggest that residents are holding back the restoration of their ecosystem. On the contrary, their opposition is, at least in part, a result of the dismissal of their localized expert knowledge by the institutionalized expertise of scientific knowledge. A third answer to the above question is that alleviating residents' anxiety and fragility *is* coastal restoration. The people are part of the ecosystem and have been in an exchange relationship with that ecosystem for hundreds of years. Yet they feel cut off from the possible recovery of their home. Ironically, perhaps, residents may be treated as just another species in the ecosystem and consulted as much as would be the birds, fish, or flora species about the restoration process. Nonetheless, quelling the anxiety and fragility of residents is an integral part not just of Louisiana's restoration but of similar small- and large-scale ecological restoration processes that will certainly take place in the future and that we are now faced with. Again, addressing the concerns of communities is part of restoration.

The in-depth, active involvement of residents in coastal restoration I am proposing in this chapter appears more imperative now, in the post-Katrina era. However, since the process of coastal restoration is something residents feel is bureaucratized and authoritative, and this appears to be increasing in the post-Katrina era, then it seems that the only way that residents can gain meaningful involvement is to demand that they be inserted into the process. It is likely that without community involvement Louisiana's coastal restoration is in grave danger because of its unsustainable nature due to bureaucracy and physical reliance on mechanized restoration, not to mention the lack of long-term political commitment. Furthermore, the degree of community involvement holds implications not only for academic and mere policy recommendations but for how we actually live in places and for what those places are like.

ALIENATION, ATTACHMENT, AND RESTORATION

Southeastern Louisiana's coastal residents conveyed a damaged sense of self. They expressed a vulnerable identity, thus their anxiety and alienation. Residents' alienation stemmed from feeling "cut off" from the restoration process. Because residents felt distanced from a process that acted on an object with which they identified, their anxiety increased. The concept of alienation is being applied here in the Marxian (1844) sense where residents are alienated, estranged, feel a certain separation from self, community, process, and product. Importantly, the feeling of alienation ensues because people get the sense that there is a separation and violation of things that naturally go together—self, community, process, and product.

In this sense, alienation arises through commodification of our work lives. In other words, we put much of our life's energy into the product of our work, an object, but due to commodification of that work, our lives no longer belong to us, but to the product of our work (Marx 1844). Residents of Louisiana's coast, less so now, but in terms of a continual line of generations, have put their lives into an object, the land, and the product of that object, the fisheries and various agricultural activities. While there are certainly commodifying features of that life, residents did and continue to form a sense of self with place where their labor and hence their products belonged to them. A signature feature of this relationship between work, self, and land/water is a sort of natural affinity between community members and the land. Coastal residents were dependent on the land, and the land, or the sustainability of it, was dependent on them. Alienation was largely absent in this process. Fishing and agricultural activities emphasized an affirmation of self where identity was marked by their work. These activities are inseparable from place, providing a connecting thread between work, place, and self.

While economically advantageous, the influx of the oil industry and other technological developments produced less direct engagement in creative and self-affirming labor by residents. If it is not obvious how work such as that in the oil industry is more commodified and thus alienating than aquacultural and agricultural activities, then it should be clear that this more industrial work turns the land, or that which is in the land, into a product that is to be acted upon, not with. No one ever put oil back into the ground because it wasn't ready yet, cultivated the field to yield more oil, or refrained from drilling to allow replenishment. However, stories of identity that are rooted in past generations of place-oriented labor, as well as other current forms of work and leisure that are linked to place, keep the self identified with and attached to the locale. This is not to romanticize the hardship of life in the past, but it does notice what is, now, because of a more commodified time, a sense of loss. As one "old-timer" from St. Bernard notes, "We used to live *off* the land, now we live *on* the land."

Thus, as land loss increased and it became increasingly clear that much of it was caused by human activity, estrangement or alienation arose. Fewer in the community were involved in the self-affirming labor of agriculture and/or fishing. Add to this decrease of interaction with the land and waterways the fact that these coastal residents, as all of us do, relate to their environment, the object, through relating to others. As a result, residents' alienation is experienced in reference to others. The degradation of the land is perceived to be mostly the fault of outsiders—mostly oil companies and politicians, but also developers, environmentalists, and governmental agencies.

Alienation is exacerbated as the restoration process is undertaken. For restoration provides another instance of an object that is being produced, not produced, or poorly produced out of something that residents get an essential part of their being from. That is, residents acquired their essential being from what Marx (1844) would call the "free and conscious activity" they engaged in with the land. Thus, unable to engage with the recovery of something that provides them with identity leaves residents more alienated. They are denied recovery-oriented interaction, activity, life-affirming labor with that which they believe gives them identity—the land.

It is the interaction—the exchange relationship—between labor and land that connects land, place, and person. As residents are denied participation in the recovery of place they are also denied the recovery of self. Consequently, residents experience a sense of estrangement or alienation from self. Thus, restoration comes to symbolize powerful others denying residents a particular sense of self. Although there was not a universal resentment of the restoration process, residents' anxiety was widespread. What is important here is that this anxiety results from feeling powerless to stop the eradication of land. Their attachment to place was constructed through sustained interaction with the coast and in this way, their anxiety begs for the opportunity to heal itself through the same processes that their attachment was made. As Jeppa said of those charged with fixing the problem, "They need to get with the people who work out there and live out there to really get a feeling on what's really happening."

While Jeppa's words call for more community involvement, further commodification of the restoration process forces alienation from community. In short, alienation from community arises because there is little communal ownership of restoration; it is a commodity produced and implemented by outsiders. Modern societal norms propose technological commodities to provide remedies for a myriad of societal and individual defects. Thus, coastal residents experience an inner struggle where some part of them wishes to reconnect with place through actively assisting in its recovery while another part of them expects the state to provide technological fixes for their ailing environment. As the urge to be an element in recovery goes unfulfilled, expectations of what the state should do create dependence, dependence by communities on outsiders and, consequently, more resentment.

Residents become estranged from community because it is the community who derives identification from the land and it is community who ails because of its degradation. Thus, not being able to participate in place and self-recovery is alienating, in part because of the dependence on others to do the work.[1] The manifestation of this dependence can be somewhat deceptive, however, because it is *expected* that others are to do the work. Alienation from community increases as the product is perceived to be shoddy and, in a more abstract sense, as residents are denied the ability to attempt to solve the problem in the very ways that they built community, together, through interaction with the land. Consequently, resentment increases and conflict arises.

Dependency on outsiders also exacerbates alienation from the restoration process as it appears that solutions can only come from specialized technical procedures. These specialized processes confront residents as an alien power that overrides their localized ways of knowing. Residents become estranged from and come into conflict with a process that fails to restore a place that they believe they have an intimate knowledge of because of their interaction, activity, and life-affirming labor with that place. On the other hand, I do not want to oversimplify the case. There are many issues not taken into account here such as a myriad of laws and regulations, the role of history as well as the current historical moment, who has the power to get their particular story of place into the public realm that, in turn, shapes community discourse, control over resources, and states' rights v. federal regulations.

However, these elements are appendages of the primary alienation process discussed here.

Finally, residents are alienated from the product, the recovery of their environment, because, in short, there is none. It is the result of alienated labor, of an alienated process. It is the ends of an alienated means and holds none of the essential nature or activity that gave the place meaning in the first place. Since this product does not even materialize, there is conflict over a product, the recovery of land, that was promised. Hence, alienation increases as residents are made dependent and denied control over their own recovery. The alienation of self, community, process, and product from a place that the self relies on for its meaning holds consequences for how proposed restoration of place proceeds.

ATTACHMENT AND RESTORATION

On the same side of the coin of this idea that environmental restoration is alienating is the notion that cultural norms lead to the damage in the first place. Clayton and Opotow (2003) point out that there is increasing recognition that environmental degradation is not only a result of technological issues, of neglectfully using resources, but a cultural issue, behavioral and attitudinal. This realization leads to focusing more attention on how people think about environments. Even so, both the natural and social sciences treat the relationship between humans and their environments as a disaffected one. This relationship is construed as an economic one instead of an affected relationship. Most research deals with issues of sustainability, with deducing how people can be convinced to "make personal sacrifices for the environment through recycling or reducing their resource use" (Clayton and Opotow 2003, p. 3).

The meaning that coastal Louisianians attribute to their experiences certainly contrasts with the view of people as having mainly economic interests in environments. Even their economic concern is affectively related as it constitutes not just occupational and economic stability, but a livelihood that is tied up with place through a generational exchange relationship.

While many modern places may be the result of a disaffected but economically interested relationship, ecological planner Eric Higgs (2003) states that, much like coastal Louisiana, "some places are significant in the lives of a community because they preserve the deeds and memories of those who came before" (p. 148). There is what he calls "narrative continuity" where these meaningful deeds and memories are passed along through generations (Higgs 2003). The residents of coastal Louisiana don't just pass along this history verbally but through a varyingly reenacted narrative. That is, significance is also transferred through *living* in the same place, whether that is by engaging in equivalent occupational practices or simply by living in the same meaningful place where similar everyday practices and rituals are engaged in by subsequent generations. Reliving the narrative is akin to returning to a favorite childhood place or vacation spot with one's own children and engaging in fondly remembered activities, thus re-creating and reinforcing significance and passing that along to the next generation. For Louisiana's coastal citizens, this "narrative continuity" is what gives place meaning beyond economic processes. It is, as we have seen, an integral part of identity.

Just as there is growing recognition that degradation of our environments is culturally entrenched, there is also the increasing realization that the meanings and saliency of places are symbiotically formed through ecological and cultural histories. Higgs (2003) argues that "the narrative continuity of a place is formed of both ecological and cultural histories; the two cannot be easily or appropriately separated, although often they are" (p. 153). The narratives of Louisiana's residents avidly convey the conflation of the natural environment with symbolic social meaning. Additionally, it is the quickening advent of land loss that heightens and forces acknowledgment of this fusing of self and ecosystem. It is this slow disaster, and more recently Hurricane Katrina and, for some, Rita, Gustav, and Ike that causes residents' heightened awareness that in relatively unthreatening times would be more in the background of consciousness.

Whether in stable or uncertain times, the meanings we hold for places are the result of social and natural histories which, in turn, inform our actions. Because there is an "ongoing dialectical relation between human acts and acts of nature, made manifest in the landscape" (Crumley 1994), then ecological restoration should account for our ecological and social histories (Higgs 2003). In the past, what we have done or haven't done to natural places has depended on the meanings we gave those places. These actions helped to shape the subsequent meanings we gave places, and this process continues today.

As we are faced with replenishing locales due to our past actions, restoration should account for the changing meanings that nature and natural places have undergone. And meanings change in different but related ways. The meanings we give to an ecosystem change due to the ongoing nature of science, cultural and economic shifts, and because of local influence, for example. Restoration projects must consider what natural places have meant in the past, what they now mean, as well as the direction that we would like meanings to take in the future. In many ways, restoration takes place because of prevalent current meanings, and, if history is ignored, then, as Higgs states, "we will be giving in to the capricious nature of contemporary judgment" (p. 131).

RESTORATION, TECHNOLOGY, AND COMMODIFICATION

At least according to these residents, current restoration in coastal Louisiana appears to be doing just that—ignoring history, especially local history, and yielding to the impulsive nature of contemporary meanings. Prior to Katrina and Rita, and now in the post-Katrina era, restoration is proposed and implemented along commodified and consumptive lines. Through the procedural beliefs that make up contemporary forms of ecological recovery, the current restoration process seeks to re-create a natural ecosystem and the processes therein through a massive mechanized system cut off from significant human engagement.

In fact, according to Higgs and many involved in the restoration movement, what is occurring within Louisiana's coast is inappropriately termed restoration. Reparation or regeneration would be more apt terms. Reparation, looking backward, would be more suitable because the process seeks "to restore and/or mimic the natural processes that built and maintained coastal Louisiana" (U.S. Army Corps of Engineers 2004). Or regeneration, which looks forward, might be a better term because the process aims "to establish highly productive, cost-effective, and long-term coastal restoration projects that are essential to saving Louisiana's coastal wetlands" (U.S. Army Corps of Engineers 2004). This idea, based on current meanings of economic productivity and efficiency, points forward to saving what exists for an ill-defined future. The rhetoric of Louisiana's coastal restoration is not wrong per se, but these ideas that constitute its primary focus further commodify the coast and leave out important meaning necessary for real restoration.

While the ideology of Louisiana's coastal reparation or regeneration is riddled with the word restoration, as are the two quotes above, the process is not restoration as it has developed and come to be defined in professional spheres both from academics and practitioners. In 1996 the Society for Ecological Restoration (SER) came up with this streamlined definition for restoration that has much accompanying text: "Ecological restoration is the process of assisting recovery of an ecosystem that has been degraded, damaged or destroyed" (Higgs 2003, p. 110). I understand that it could be argued that Louisiana's restoration is "assisting recovery." However, the U.S. National Research Council's (NRC) often-cited definition states that "merely recreating the form without the functions, or the functions in an artificial configuration bearing little resemblance to a natural resource, does not constitute restoration" (1992). Indeed, Louisiana's restoration only artificially re-creates form and function. Additionally, history and what meaning we are creating don't seem to be seriously considered. In restoring Louisiana's coast, do we indefinitely deliver "slurry" dredged by enormous scouring barges and pumped through pipelines? Do we just keep building bigger and newer river diversion projects? Do we employ this style in other places as well?

Currently, freshwater and sediment diversion projects constitute the majority of Louisiana's restoration. These are man-made technical structures built under contract of the U.S. Army Corps of Engineers. They require a technical expertise, and, as we have seen, residents feel intimately imposed upon by the projects and the accompanying ideology. There are more localized endeavors like revegetation projects; however, it appears that the institutionalized technical expertise of such practices has excluded communities who communicate a longing for restoration.

Louisiana's coastal restoration is technologically based, and this reflects the technological dependence of contemporary life. Higgs (2003), citing the U.S. philosopher Albert Borgmann, sees much of technology as "a distinctive pattern wherein we displace things and activities that matter to us in favor of commodities" (p. 180). We should be concerned, as Higgs warns, that because of our dependence and the resultant expectations by citizens upon the state and large corporations, ecosystems and restoration will be converted into mere products. In today's climate, partnerships between the state and corporations to support restoration initiatives further commodify and create dependence on these institutions not only for the economic, but now the environmental well-being of our communities. Corporations such as Shell Oil Co., the primary sponsor of the state of Louisiana's media campaign, "America's WETLAND: Campaign to Save Coastal Louisiana" (http://www.americaswetland.com/), put on an ecologically concerned public face while they spend millions in lobbying governments to weaken the restrictions on the lands they claim to be concerned about.

Because of this collusion globally, many experts point to Louisiana's depleting coast as a sort of canary in the coal mine of more large-scale ecosystem failures to come. It certainly appears that, thus far, Louisiana's restoration holds something further to warn us about. This rehabilitation has turned into a technological commodity where we are reliant on a specialized professional niche to produce a good. Through specialized and mechanized remediation processes we turn restoration into commodities that can be bought and sold. Nowhere is this more apparent than when projects are bought and sold through competitive bidding processes by engineering firms. Communities become mere spectators. To put the effects of commodification another way, which type of health care works best: the type where the patient is passive and submissive to the authority of the expert, the doctor, or the type in which patients are active participants in the care of their bodies both by lending their own expertise and by physically engaging in their own care?

Just as engaging in our own healing yields a higher recovery rate and is an invaluable gift we can give ourselves, participating in the recovery of place can generate similar rewards. While commodities appear to be a dominant force in contemporary social life, there are meanings that cannot be reproduced by commodification: "the camaraderie of a Saturday morning stream cleanup, the knowledge that a flourishing meadow involved some of your own labor, the experience of seeing a new species of bird begin to inhabit a restored wetland" (Higgs 2003, p. 180), or participating in the planting of new cypress trees and then watching them grow from saplings into mature trees in what was once a desolate area. The meanings generated from these actions rely on experience and resist commodification.

Certainly, the meanings that coastal residents give to place go beyond products that can be exchanged. In fact, that is what they intend for us to understand—that place cannot simply be quantified in economic terms. In many ways, it is invaluable. Thus, conveying such intimate meanings serves as a call for the salvation of place and thus the preservation of self.

Selves and communities are alienated by Louisiana's coastal restoration because, being commodified, the context is stripped from the process, "leaving machinery and a commodity—or mere means and mere ends" (Higgs

2003, p. 185). Communities used to engaging with their environments have had it transformed into a product. This seems to have occurred almost imperceptibly to residents, perhaps making the restorative process all the more perplexing. The very way residents reify and make meaning with place is through their entrenched engagement. But, as we have moved toward a more technical and commodity-driven society, we come to rely on product instead of process. This is cultural. Technology has become a system, a mindset. We have approached the point of "technological saturation. The more pervasive something is, paradoxically, the more it is concealed. We tend not to notice what is all around us" (Higgs 2003, p. 190).

In general, we know less about how things work and are primarily concerned with ends. So it follows that if environments are damaged then we need the business of government to produce a solution. Perhaps we as a society, as well as coastal residents, have been lulled out of active participation with our own environments. Engaging our environments may seem even more daunting if degradation is immense and caused by unidentifiable others such as storms and economic development. Thus, not knowing how such a large ecosystem can be restored, residents turn to the state, only to be alienated because of the technical commodification process that is itself a product of the economic forces that drive and sustain the state.

In the face of what appears an overwhelming problem, residents feel more anxiety and alienation because the meaning generating practices that built place attachment in the past have now been stripped away. Yes, they still engage in their normative processes and activities, but residents watch the deterioration of their communities while not being able to engage that decimation. These people, like those of many rural communities, have traditionally engaged environmental problems themselves. Yet here they feel helpless. Engaging their environments was how place was given significance; it is how place became identity. "It is this communion between self, thing and environment (and perhaps also spirit) that generates profound meaning in our lives" (Higgs 2003, p. 185). This communion is made up of what Higgs calls "focal practices," ordinary experiences such as celebratory meals, gardening, fishing, walking, or even shoring up one's land, throwing back fish that aren't mature or big enough in order for the whole population to thrive, or selective thinning of dense forest undergrowth to allow for a more resilient and healthy ecosystem.[2]

Focal practices give our lives meaning. For Louisiana's coastal citizens, focal practices are how they have traditionally gone about life. This disaster invades those practices. Consequently, the loss of land is salient among residents. Watching this incessant encroachment while not being able to focally address this threat produces personal fragility, subsequent anxiety, and then estrangement, resentment, and conflict with the technical restoration process. Interestingly, Hurricanes Katrina and Rita, while causing widespread destruction, provided opportunity for focal practices. Shortly after Katrina, Terrebonne residents who suffered significant damage but not total devastation like those of St. Bernard, Plaquemines, and Lake Catherine were "already starting to rebuild." As Kimberly Solet, a reporter for the *Houma Courier*, conveyed to me the resilience of the residents, it became clear that those who could were engaging in focal practices that reified their meanings of place, something that they had become estranged from over the preceding years.

While the storms may provide some with an opportunity to reclaim the place attachment that was threatened by those disasters, coastal restoration does not. Restoration processes are not supposed to be alienating. Ecological restoration should be a focal practice; however, it is not inherently so. It has to be made so. As it is conceived and implemented in projects both large and small, it is meant to build and give new meaning to place, incorporate cultural practices, and employ rigorous scientific research and application (Higgs 2003). While it is not limited to these criteria, the argument could be made that Louisiana's coastal remediation applies these elements. It is building and applying new meaning to place, although probably not the meaning that was intended. It incorporates (meta)cultural practices as it reflects a technologically commodified culture. And it definitely applies rigorous scientific endeavor, yet is cut off from the community and society which it serves.

Louisiana's remediation may be focal for the scientists, engineers, and other specialized professionals who are engaged. However, even they are severed from the meaning that is meant to be generated by ecological restoration. The competitive, politicized, and bureaucratized nature of the process is likely to leave many dedicated natural scientists frustrated and alienated from the connections that would emerge from assisting the recovery of the ecosystem. Furthermore, genuinely motivated professionals are likely to be resentful of residents who they may feel are unappreciative of their efforts.

Even if Louisiana's coastal remediation is successful, what meaning of place will be generated? How will the meaning of place be transformed for all? What will engineers', scientists', and residents' symbolic landscapes look like? What will these new definitions mean for place and how we use it? Does this constitute successful restoration?

The answers to these questions, if the remediation continues along its current path, are likely to be riddled with conflict that sequesters environment and place into a perpetually unsustainable state. The aim of ecological restoration is not alienation and resentment by all involved parties. Focal practice is an integral part of the process, but only if we direct the practice toward valuing environments and the social relations that form and exist within the midst of restoration (Higgs 2003).

The goal of restoration should be both ecological and cultural. This goal has greater importance when restoration is being attempted in a place like coastal Louisiana where there is a large culture that has formed around and in concert with that ecosystem. And while the voices and research of scholars and scientists can add authority to the idea, it doesn't necessarily take academics to point out that participating in restoring one's own community creates independence and self-worth, and builds connections to place, both individually and communally. The focal practices of some Egyptian residents and their own experiences with coastal erosion provide a partial parallel to Louisiana's.

Rising sea levels in the Mediterranean Sea due to global warming are causing saltwater intrusion up and into the Nile River. Much like many of the residents of New Orleans, for whom the effects of coastal land loss didn't register until Hurricane Katrina, the citizens of the coastal city of Alexandria are mostly unaware of their erosion. Nonetheless, rural farmers notice saltwater intrusion into the groundwater and its effects on growing crops. Salah Soliman of the University of Alexandria says that, just like the Mississippi River, the Nile used to deposit soil throughout the region until the Aswan High Dam cut off that annual gift in 1970. Consequently, erosion and rising seas send salt water inland, increasingly contaminating soil as well as threatening a major highway used for commerce. Salah Soliman predicts the highway will be under water within ten years (Hansen 2008a, NPR).

Yet, up the Nile from Alexandria in nearby Cairo, the people of the poor district of Manshiyet Nasser are engaging in focal practices and bringing about economic and environmental sustainability. In concert with some young, innovative environmentalists like Thomas Taha Rassam Culhane, a doctoral student in urban planning at UCLA and founder of the nongovernmental organization Solar Cities, the citizens of this poor urban district are installing solar hot-water heaters on the rooftops of homes and apartment buildings in Cairo's slums. One of these solar water heaters sits atop their informal neighborhood recycling school, which was developed by UNESCO with Proctor & Gamble in order to encourage more technical skills and a community business strategy from the community's decades-long practice of recycling garbage by hand. Here, young people learn about the economics of recycling. Supplied by their solar energy, they also learn "to shred plastic in machines, wash and dry it, bag it and send it out to be melted down for reuse" (Hansen 2008b, NPR). Culhane says that a "collective intelligence" is being fostered through a "participatory process" where "local steel cutters, copper welders, and glass makers" add their expertise in constructing the water heaters (Hansen 2008b, NPR). The focal practices and sustainable community building by these residents are especially important since it is predicted that rising seas will create millions of environmental refugees in the region over the next century (Hansen 2008b, NPR).

Focal practices and creating community independence through sustainability is no longer an exotic idea reserved for the exceptional few. Majora Carter's Sustainable South Bronx (http://www.ssbx.org/) is an urban example of breaking the chains of economic dependence and ecological injustice by green skills training and education where the business of green roof installation has been forged. Fair Trade coffee and sugar production continues to grow even as farmers face challenges due to increasing demand for their products (Jaffe 2007).[3] Urban neighborhood tree-planting and maintenance projects (Austin and Kaplan 2003), communal gardening in domestic violence shelters (Stuart 2005), and school-based programs that employ adolescents and elementary students as primary actors in the management of local public lands, wildlife sanctuaries, and the creation of wildlife exhibits (Thomashow 2005) are just a few other ways that focal practices are fostering community self-reliance. In fact, because of the recent tremendous growth of community gardens, urban agriculture, and the like, the reader of this will also likely know of many examples that are bringing community and self-reliance back into our modern way of life.

Indeed, the seeds for the nourishment of this type of practice already exist in coastal Louisiana; they highly value place. To engage in its recovery and thus the recovery of self will further enhance and reform the value of place. According to Higgs (2003), community participation in restoration is what gives the practice its unique character, setting it apart from other practices like reparation or regeneration.

The importance of community participation and the expert knowledge residents bring with them to restoration is not meant to devalue the contribution of science to the practice. Sustainable restoration cannot succeed without science just as it cannot succeed without community and cultural involvement. Science analyzes and achieves prediction amazingly well. And analysis and prediction are essential to the success of restoration. However, it has been increasingly recognized that science alone cannot substitute for knowledge that has developed over time by those who are attuned to the land in ways that science is not. This knowledge, known as traditional ecological knowledge and wisdom or TEK and TEKW, is based on struggles for identity and survival by traditional cultures (Higgs 2003). This type of knowledge is organized around oral tradition and passed from generation to generation and is its own type of focal practice. Louisiana's coastal residents know they possess this knowledge and feel it is being dismissed as lacking any value. So, in order to value community knowledge and participation as part of ecological restoration there must be active engagement by community in the scientific and engineering endeavor.

PARTICIPATORY RESEARCH AND COASTAL RESTORATION

As Higgs (2003) simply states, "[R]estoration is doing well when it nourishes nature *and* culture" (p. 226). When residents participate fully in restoration, value is added and community is strengthened. Attachment to places like coastal Louisiana is enhanced and reified, and new insight is gained (through engaging in scientific practice). What Higgs and others, including many in the Deep Ecology movement (Merchant 2005),[4] are calling for is for communities to participate in the restoration of ecological places for self-empowerment. Primary to this transformation is research dedicated to the social good, public participation, and "the incorporation of humane values into research goals," not just disciplinary inquiries (Merchant 2005, p. 112). Participatory research, heretofore largely relegated to the realm of social research, fulfills these requirements.

Participatory research (PR) is a social and academic movement that has gained momentum over the last twenty years. It may be exactly what ecological restorationists are calling for. However, PR goes further than Higgs's (2003) focal practices by adding the element of community involvement in the actual research.

Participatory research has two basic characteristics: 1) increasing community participation in the research process and 2) producing social change (Stoecker and Bonacich 1992). Bringing PR into ecological restoration would enhance social change through assisting in the recovery of ecosystems. Participatory research revolves around issues of social justice, and, in a place like coastal Louisiana, just as we are increasingly noticing in most places, social and ecosystem justice share similar goals. Indeed, what many are finally realizing is that our environmental problems and our social problems are intertwined and they can be, or perhaps must be, alleviated singularly.

In fact, PR requires that communities become such a part of the scientific endeavor that even the research questions, and thus what is studied, comes from them (Stoecker 1997). Traditionally, it is the professional, scientist, or academic who poses the research question. However, as practitioner Randy Stoecker (2005) states, in PR, the research question must always come from and be defined by the community. Following this line of thought, an automatic response becomes "How can communities come up with research questions that will address complex problems that will, in turn, lead to the repair of damaged ecosystems?" Well, it isn't hard to imagine that if the residents of coastal Louisiana were asked twenty years ago to address an issue for study that they would have come up with a research question that sought to solve the loss of land in their environment. Then it becomes the role of scientists to assist the community, through sharing their specialized knowledge, in refining that research question into subsequent, more specific questions while incorporating community knowledge. Scientists lend their expertise at the service of the community (Stoecker 2005).

Thus, democratizing the knowledge process is essential to the practice of PR (Stoecker 2005). Because PR usually takes place in the realm of social science this has meant that local knowledge may sometimes be the object or subject of research but almost never part of the research. Within PR local knowledge becomes part of the theoretical and methodological conversations providing input into how research should proceed. Additionally, community members gain research expertise from professional researchers or academics and, likewise, the professionals acquire knowledge from the community. In this way, the hierarchical relationship between community, science, and even the state is dissolved. And not only does the research question come from members of the community, but they are actively engaged in other methodological procedures such as data collection, analysis, and dissemination of results. For Louisiana, this would mean that coastal restoration would have to be retroactive, but it is never too late. In fact, just as the current bureaucratic and commodified form of remediation may have been further institutionalized by the storms of late summer 2005, the current historical moment also provides opportunity for reforming the process into a more focal restoration.

To be sure, this is a lofty goal, but to achieve ecological restoration this is what must occur. The conflicts that are pervasive in the current restoration process in Louisiana will prevent the engineers, biologists, and other natural scientists from implementing a plan that will effectively restore land, abate land loss, or be sustainable. Furthermore, prior to Hurricanes Katrina and Rita political will was not sufficient to fully fund restoration efforts or to successfully regulate the region from destructive processes. Although in the post-Katrina era there is much lip service heaped upon the immediate need for restoring the coast, it will not get the funding that most say is needed. There are too many national and global issues that are also politically pressing. Oil and gas production royalties paid to Louisiana for offshore oil and gas acquired from its waters won't do the full job either. Even if the state finds half of the billions it needs, will we indefinitely engage in the current forms of resource intensive restoration projects mentioned earlier? The continuous supply of money and resources that this would require is not sustainable.[5] While there must be significant funding for the coming decades, this reality provides further impetus for community participation.

Regarding communities and their knowledge being part of research, we can certainly envision residents doing more than planting some marsh grass or a few cypress trees. Their knowledge of lost land formations, waterways, and aquatic life can be helpful in theory modeling. Community knowledge can also be useful in collecting data. The combination of local expertise along with scientific knowledge can be helpful in deciding the best locations from which to collect data, as in water and soil sampling, and places where loss of land varies, for instance.

In the case of analysis, local knowledge would be meaningful for interpretation. And who better to disseminate the results of studies than residents? They would possess ownership of knowledge from studies that they participated in and would spread these results throughout their communities, by and large trusted by their peers. In turn, this lends some amount of credibility to the researchers. Also gained is added political clout for communities, researchers, and restoration projects. Additionally, community-directed studies can lead to clearer regulation of wetlands and the coastal zone while also assisting communities in developing small-scale practices of sustainable restoration. However, implementing PR in Louisiana would be an arduous and almost certainly painful process, but the gains for the ecosystem, and here I include the human communities, are immeasurable.

While the practice of PR would go a long way in deflating the commodification balloon, commodification continues to gain ground in ecological restoration efforts both nationally and globally. In this climate, government agencies are less likely to sponsor longer, more unpredictable restoration processes where communities must be incorporated into and conduct research. A professional firm offers more guarantees. It is understandable that in the commodified climate we have created for ourselves that, as monies are allocated, professionalized restoration is more aggressively sought because the national public, viewing the issue through a national and political lens, will want assurances that the coast will be "fixed."

As a result, the U.S. Army Corps of Engineers is unlikely to actively pursue PR as a research methodology in Louisiana's coastal restoration or in any of its other endeavors, for that matter. Considering Louisiana's coastal residents' identification with and deep attachment to the region, they must insert themselves into the research and restoration of their environment. Otherwise they risk continued cognitive and emotional displacement, which appears to many of them to be a prelude to their eventual physical displacement from their home.

Yet, even if residents succeed in convincing governments to make them a part of their homes' restoration, there is the risk of only symbolic participation. As efforts to combine different U.S. government offices under the title of Homeland Security attest, government agencies and institutions are resistant to change. Thus, there is the danger of PR being promoted unreflectively with continued resistance to welcoming outsiders into the institutional structure. So, even if we continue with the speculation that Louisiana's coastal residents are able to gain entrée into restoration, the struggle to create a sustainable and independent place would be long from over. If community participation is not accompanied by reflective theoretical and methodological understanding, then participants will be more disempowered and alienated than prior to their participation. For example, if residents are collecting water samples but have no theoretical or methodological understanding of why they are doing what they are doing, then there will be further conflict and the ultimate breakdown of the process.

Furthermore, if residents did not gain adequate understanding of theory and method—why take water samples and what that means for land loss—then, in addition to being denied necessary self-sufficiency, they could not aptly apply their own expert knowledge to the process. In turn, this would provide seemingly substantial evidence to the institutional experts that local knowledge is of no value. Participation should not be a gift that powerful outsiders grant to powerless insiders, but the powerful may certainly come to view it that way.

On this matter of the public and restoration, engineers and biologists working in coastal Louisiana may wonder what they have to do with social change and what that has to do with restoring the coastal ecosystem. The voices in this study and the conflict and frustration scientists meet within communities should provide the answer. The good news is that more and more scientists are realizing this and continually calling for an interdisciplinary and public-oriented approach. As the well-established coastal Louisiana biologist Robert Twilley has said, "The social sciences have to be at the forefront here to establish what exactly are the consequences in every restoration decision and we have to have an honest dialogue about that" (Joyce 2005). Indeed, there are numerous social scientists involved in restoration efforts, but this is difficult work and the idea of integrated community involvement is still, as we say in south Louisiana, lagniappe.

It must be obvious that if the ecosystem is to be restored then social change must occur. Enabling social change means providing a realm for the humans who occupy that ecosystem to repair themselves and their environments. Hence social change must occur within the scientific community as well as within the public sphere.

Just as ecological restorationists know that successful and sustainable restoration involves community participation, historical fidelity, ecological integrity, and allowing for change (Higgs 2003), PR acknowledges that positive social change results from a shared and productive knowledge between scientists and communities. Participatory research produces knowledge that is contextual, knowledge that is shaped by local conditions, knowledge that assumes a changing dynamic between community, culture, environment, and even science (Stoecker 2005, Cronon 1983). It assumes that these interactions are dialectical.

Something called "local science" is what comes from the dialogue of PR (Couch, Kroll-Smith, and Kindler 2000). Local science consists of the insights of local people, communities, and institutions. It does not replace scientific insight but is combined with it, producing understanding and expertise of patterns shaped within a local context. Furthermore, valuing and incorporating local wisdom into research does not mean wholesale acceptance of that local intelligence. However, it does take skill to respectfully question and to take into account what, how, and by whom particular knowledge is produced (Burley, Couch, and Kroll-Smith). This *is* the dialectic.

One of the strengths of the researcher is the expertise of critical thought. However, in PR this is done in concert with the valuing of local knowledge, with the extracting of useful and applicable local insight, while community members simultaneously acquire this proficiency. This skill is a new and difficult endeavor for social scientists to acquire, not to mention natural and physical scientists for whom this will undoubtedly be a task. Thus, restoration processes like coastal Louisiana's must incorporate individuals that have some PR experience.

The goal is empowerment of the ecosystem and its human inhabitants so that both can thrive. Participatory research realizes this goal and its relationships. Participatory research will be empowering if researcher knowledge and local knowledge are mutually valued and combined. Stoecker (1997) believes that attaining this requires two elements. First, participation must consist of a "cogenerative dialogue"[6] that derives data and perspectives from multiple experiential and expert models. Second, communities must be provided the opportunity to, at the very least, collaboratively control the decision-making process (Stoecker 1997). It is worth noting that a by-product of good PR would be a bridging of the gap between science and the public and also a reengaging of youth in the passion of science, something that many have said is deeply needed. These goals, skills, and by-products must be initiated and guided by the researchers, whose role is "to foster participation through modeling inclusiveness, valuing diversity, being attentive and responsive to community issues, acknowledging strengths, achievements, and contributions by community members, and by building a consensus building approach to decision making" (Burley, Crouch, and Kroll-Smith). By employing PR, those involved in Louisiana's restoration, and restoration projects in general, can nurture nature *and* culture. For as Louisiana shows us, and as is becoming more and more clear, nature *is* culture.

CONCLUSION

For all of the benefits it can bring, there is much cause for skepticism about PR being implemented into Louisiana's coastal restoration. With possibly billions of dollars on the line, it is hard to see that state or federal governments will allocate money for unpredictable, tenuous projects where local communities are directly involved in decision making and participating in what is traditionally a positivistic scientific endeavor. Timelines would be vague and open to change and the possibility of failure would be construed by many as high.

However, we must consider what it is that is at stake. First, are current ecological remediation efforts sustainable? Can human technological and mechanized processes create ecosystems that will function far into the future? Second, considering the state of the endeavors thus far in Louisiana's coastal zone, can remediation plans succeed in implementation? Are these efforts too alienated from the human communities? Does it decrease their autonomy and increase their economic and ecological dependence? If so, does this alienation and dependence create conflict that impedes remediation efforts?

The voices of residents appear to be calling for the salvation of themselves and their environment. They feel helpless and confused. It appears that the way they can reclaim place and self is through PR. However, it is almost certain that this will be welcomed by few. Therefore they must insert themselves into the restoration process. This will mean organizing and acting regionally as a whole, as well as more locally in individual communities. No doubt this will be a difficult and often painful process. But as the climate changes and environmental disasters become larger and more frequent, sizeable, government-funded restoration projects will become more prevalent. And this means that it is likely that future restoration projects will face the same challenges that Louisianans face, from commodified and mechanized restoration to political posturing to the ruination of the human communities within the ecosystem.

In coastal Louisiana the self that is defined by its relation to place plays a dominant role in overall identity. In addition, I would venture that place plays a much larger role in identity there than in more modern, suburban places. At the same time, there are many regions and locations around the globe where community and culture is as entrenched in place as it is in coastal Louisiana. And the extent of Louisiana's environmental degradation is certainly not unique to the region. Do we also deny them restoration, as Thomas and Mayheart Dardar allude to in their poignant poem in the afterword? Lastly, the loss of Louisiana's coast is not just a lamentable loss that is somehow part of the natural order of life and death. The loss of these places has an impact on all of our lives and all of our ecosystems. Are we all expendable?

AFTERWORD
The Path Ahead

At no time in the memory of the current tribal population have the Houma people faced a trial such as the one confronting us today.

In recent years four major hurricanes have impacted the United Houma Nation, Katrina and Rita in 2005 and Gustav and Ike in 2008. In both instances they hit within weeks of each other, back-to-back blows that have challenged the strength and tenacity of the Houma people.

Hurricanes, in themselves, are not a new obstacle to the tribe. We are coastal people who have lived in south Louisiana for centuries. There are those who wonder why we live in such a "vulnerable place." The answer, of course, is quite simple; our vulnerability comes from a century of unchecked development that has swallowed the natural defenses that once protected us.

So now, in the midst of our current recovery effort, we take a breath and look to the future. What will it take for us to survive as an indigenous nation as the avarice of empire continues to devour our land? For decades now we have watched the effects of coastal erosion as over thirty square miles of Louisiana coastland goes under the waves every year. As politicians and scientists continue to study the problem we continue to wash away.

What we have recently learned, above all other lessons, is that the government will not be there for us with the solutions we need. So for the United Houma Nation to survive we must find our own path ahead.

The path ahead can be summed up in three distinct phases, three efforts that we can undertake to assure the survival of our nation.

1) Sustainability

For over two hundred years the Houma people have been identified with several distinct settlements along the bayous below the modern city of Houma. Bayou DuLarge, Dulac-Grand Caillou, Montegut, Pointe au Chene, Isle Jean Charles, and Golden Meadow have always held the majority of the tribe's population. After the storms of 2005 many homes in these places were elevated, but we have now learned that we need a more substantial effort.

A large portion of the tribe's population is determined to make their stand in these communities. Given our history of dealing with land speculators and oil companies the desire to hang on to what land we have left is understandable. As a people we stand united on this issue and are committed to make this possible.

To accomplish this our people must have elevated, hardened homes that sit above flood levels and can better withstand wind gust.

2) Security

At the same time there is another portion of the tribe's population that would prefer to establish a more secure home. Added to this is the realization that continued inaction by the government on coastal restoration will eventually force some of those who wish to remain where they are to leave. For these reasons the United Houma Nation has begun to plan for a community development project north of the Intracoastal Waterway. A new community will give the Houma people a place of refuge to gather instead of having to disperse into the surrounding nonnative communities. This area could also serve as a temporary shelter for those citizens in the lower bayous who are displaced by storms.

3) Shelter

There is also a need for a facility to shelter Houma people in the event of a major storm that forces evacuees north, above I-10.

Currently the practical solution will be a cooperative agreement with another tribe such as the Mississippi Choctaw, but there are limitations on the number of citizens that can be sheltered.

A more permanent solution would be a campground/retreat that would be owned or leased by the tribe and that would facilitate a more unified evacuation.

Understandably, these are daunting challenges, but they represent a path to a substantive future. This is a coherent vision for Houma people, a hope for a better day.

T. Mayheart Dardar and Thomas Dardar
Council Members, United Houma Nation

ODE TO ISLE DE JEAN CHARLES

How do you replace paradise?
Tell me if you have a clue.
When it finally sinks beneath the waves.
Tell me what is left to do.

For to us land is life.
A part of us as flesh and bone.
To see it gone, so hard to bear.
To stand without, cold, alone.

We've stood this earth.
We've made the fight.
Still, it has been taken.
Still, it passes from sight.

How do you replace paradise?
Can you have again what you've lost?
Can you build again in another place?
And if you can what is the cost?
Can we carry with us the grains of soil?
From which our lives have sprung?

Can we journey on as a people?
Can we sing again the songs we've sung?

I know we must keep hope alive.
We must walk on despite the loss.
A people must stand strong.
Together, there is another river to cross.

But how do you replace paradise?
How do you make a memory real?
How can your children see a place
That they can no longer touch or feel?

T. Mayheart Dardar and Thomas Dardar
Council Members, United Houma Nation

APPENDIX
General Data, the Interview Guide and Methodology

Location in Narrative Where Respondents First Introduced Land Loss				
	First Portion	Second Portion	Last Portion	Not at All
Grand Isle	10	3	5	2
St. Bernard	16	7	2	5
Terrebonne	13	8	2	0
Plaquemines	2	5	4	9
Lake Catherine	7	6	3	2
Delcambre	3	3	1	8
Totals	51	32	17	26

Table 1: Frequency of where respondents first brought up coastal land loss in their narratives.

Coastal Land Loss as a Running Theme in Place Narrative				
	1st and 2nd	1st and 3rd	2nd and 3rd	All 3 Parts
Grand Isle	1	5	3	4
St. Bernard	1	2	3	12
Terrebonne	2	0	6	10
Plaquemines	0	0	4	3
Lake Catherine	1	0	3	6
Delcambre	0	0	3	1
Totals	5	7	22	36

Table 2: Frequency of respondents who discussed coastal land loss across all major portions of the interview.

Appendix

INTERVIEW GUIDE, CONSTRUCTED BY DR. PAMELA JENKINS IN COLLABORATION WITH THE RESEARCH TEAM

1. Where did you grow up? Can you tell me about that?

2. What did your parents do for a living when you were young? Can you talk some about that?

3. Did you know your grandparents well? What were they like? What did they do?

4. Where did you go to school (elementary, high school)? What were those years like?

5. What was the first job you had? Talk about life since then.
    b. Are you married? Do you have children? Tell me about that.
    c. What do you like to do when you aren't working?

6. What places are important to you?
    a. What do these places mean to you? Why are they important?

7. Tell me about the first hurricane you remember.
    a. Did you know the storm was coming? How did you prepare?
    b. Talk about what happened during the storm.
    c. Talk about what happened after the storm. What was recovery like?

8. What hurricane sticks out most in your mind? How old were you and where were you? Tell me about that.
    a. Preparation for you? Community?
    b. Talk about what happened during the storm.
    c. Talk about what happened after the storm. What was recovery like?
       (Is it different from the storm they remember most as an adult?)

9. How has this place changed over your life?
    a. Roads (development)?
    b. Oil industry?
    c. Fishing industry?
    d. Tourism (if applicable)?

10. Are there any other ways (name of community/place) has changed physically? Tell me about that.

11. What are your hopes and dreams for (name of community/place)?
    a. For yourself?
    b. For your family?

12. What do you think you have learned in your life that has stayed with you?

SUMMATION OF GENERAL DATA

Providing some specific details can illuminate the importance of land loss to residents and add to the data in the tables presented above. The interview, as it was constructed and disseminated, can be broken down into three parts. In 126 interviews, land loss was introduced and discussed by residents 51 times (40 percent) during the first part of their narrative when they were asked to talk about personal and family history (see Table 1 above). During the second portion, when residents discussed personally important place(s) and their experience with storms, land loss was brought up (for the first time) in 32 interviews (25 percent). The final third of the interview addresses changes to place and hopes for the future. Land loss was approached in this final portion by 17 (13 percent) of the respondents. During constructing the interview, we believed that if respondents did not mention land loss up to this point in the interview, then it would be somewhat unlikely that they would bring it up during this last portion which, asking about changes to place, implies more physical aspects of their communities. The low number, 17 out of 126, as compared with the first two portions, appears to be indicative of this assertion.

Again, out of 126 interviews, only 26 respondents, or 21 percent, failed to raise the issue of land loss. Seventeen of these respondents came from Plaquemines Parish (9 respondents) and the community of Delcambre (8 respondents), which is located in southcentral/western Louisiana and at the time of the interviews, pre-Katrina/Rita, experienced significantly less land loss in comparison to southeastern Louisiana. As for Plaquemines, they were on the front lines of land loss. I did note a sense of distrust of fellow residents in Plaquemines, and it was more difficult to find interviewees there than in the other locales. When respondents were asked if they knew of other possible interviewees, some said they knew no one or no one "worth" interviewing. The distance between community members may have been due to the divisive oyster bed lawsuit discussed in chapter three, as well as other factors such as an immigrant Asian population participating in the fisheries. One Asian resident noted the resentment he felt from the majority white community. While it's a purely speculative idea, these reasons may have played a role in why land loss was not as prominent in the Plaquemines interviews as it was in the other communities.

Notwithstanding, for residents of the other communities, land loss was prominent. Forty-six of 91 residents, or 51 percent, from Grand Isle, Terrebonne, St. Bernard, and Lake Catherine introduced land loss during the first part of their narrative and only 9 of 91, or 10 percent, didn't bring up the issue at all.

In fact, land loss was so significant that it constituted a consistent or pervasive theme throughout 36 interviews.[1] In other words, it was a kind of "running theme" (Table 2 above), where residents brought up the issue in all three portions of their narrative. Along with the pervasiveness of land loss throughout narratives, another 34 residents discussed the issue at length during at least two portions of their narratives. Residents, again and again, continuously discussed the issue when talking about their childhoods, their occupations, their homes, storms, development, recreation at fishing camps and favorite fishing spots, as well as when talking about science, policy, and politics. In all, 70 of 100 people who broached land loss discussed the issue through *at least* two-thirds of their narratives.

THE RESEARCH QUESTIONS

The research questions for this project reflected the intent to explore the nature of life in a changing coast and people's attachment to that place. The questions guiding this research were:

- Which elements best characterize respondents' symbolic landscapes?
- How is change to place expressed and understood?
- What role does coastal land loss play in respondents' narratives?

The questions that informed this inquiry were phenomenological, meaning that they focused on how individuals interpreted their experiences. The first research question addressed the nature of respondents' symbolic landscapes. The interview guide, reflecting the research questions, established what the environment means to the narrators by asking questions about personal history in place. Residents were asked about childhood, family life, work, and school. One item asked people to talk about what places they considered important. They were then asked to explain why the named place(s) were important. These questions encouraged residents to contextualize the physical within an intimate narrative of place. Subsequently, in the analysis of these narratives, any passages that entailed discussion of the physical elements of place were viewed as contributing to residents' symbolic landscapes.

The second research question addressed how community members understood change. A series of items in the final portion of the interview guide requested residents to speak about changes to place as they saw them. The first item in the sequence was open-ended and asked what changes they have observed since living there. Subsequent items addressed explanations of changes in the oil industry, fishing industry, and the physical landscape, whatever they may have taken that to mean. This last item was as close as the interview guide came to directly asking about coastal land loss. How residents framed change was key to understanding how they viewed change.

Framing was also elemental to residents' symbolic landscapes because frames attach particular meaning to the changes they perceived. The interview guide asked residents to express subjective meaning about specific and nonspecific changes through the framework of symbolic landscapes. Again, since people were not asked about coastal land loss these questions were intended to give residents the phenomenological freedom to explore the issue of coastal land loss on their own terms (Smith 2004). In short, the aim was to obtain the symbolic meanings given to the physical elements of change.

The third research question condensed the first two into the more specific interest of this study—residents' conception of coastal land loss. The open-ended nature of the interview questions coupled with not directly asking residents about coastal land loss lent a high degree of validity to the phenomenological methodology of this study. The salience of coastal land loss to place attachment was established here; that is, "Where in their narratives did residents broach the subject? Did they bring up the issue at all?"

The residents' approaching land loss at the beginning of their narrative, when talking about childhood, was an indication of some significance of the issue. If land loss was left unaddressed until the latter part of the interview, when people were speaking about physical changes to place, then the issue was considered to hold much less importance for the individual. Where coastal land loss was brought up in narratives, how it was framed, context, and amount of time given to the issue (was land loss a running theme throughout someone's narrative or did he or she bring up the subject once and then forget about it?) all coalesced to reveal the role the issue played in residents' narratives of place, as well as the relationship between land loss and place attachment. Inevitably, this occurred within the concept of symbolic landscapes where the symbolic meaning given to land loss revealed the significance it had for residents' place attachment.

PHENOMENOLOGY, NARRATIVES, AND PLACE ATTACHMENT

The methodological perspective was phenomenological. Phenomenology focuses on individuals' interpretations of their experiences and the spotlight is on the knowledge of the subject. A primary way we as societal members transmit *our* knowledge and meaning is through telling stories, and we do this through the routine telling of our daily lives in casual conversation. This everyday way of telling stories is called narratives (Denzin 1989). Norman K. Denzin (1989) calls this "personal experience narratives" (p. 43), and it is in this way that the concept was employed here.

Phenomenology, based in personal experience but not necessarily in individuals, is concerned with how societal members continually interpret their social order and thus reproduce and construct knowledge (Smith 2004; Gubrium and Holstein 1997; Creswell 1997). As such, the aim was to gain insight into how subjects experience and understand coastal land loss. Since land loss is a place-embedded process, phenomenology, which suggests that our perceptions and interpretations give place meaning, was the best means to understanding how residents view land loss issues.

What became relevant was both conveyed and found in the clusters of words people deployed to situate their bodies and selves in the story of coastal land loss. In narrative form, interpretations involve characters that are portrayed in a particular fashion, oriented to a type of structure (drama, tragedy, suspense, humor, etc.), and usually attempt to convey a lesson or moral (Shanahan 1999, p. 407). Denzin (1989) even suggests that "every narrative contains a reason or set of justifications for its telling" (p. 41). In short, there is a point to telling a story. Stories, even anecdotal stories of everyday experiences, say something about who the storyteller is. The point of a narrative communicates what the storyteller feels is important and by so doing reveals elements of the teller's identity.

The phenomenological method employed here encouraged residents to tell their stories when they were asked about their common, everyday experience of place. They then subjectively constructed narratives that reflected the narrators' similarities and differences to others through the very telling of their anecdotal, everyday experiences. And by telling stories about place, respondents revealed the meanings that made up their symbolic landscapes: they transformed the physical environment into symbolic environments through self-definitions (Shanahan 1999; Cantrill 1998; Greider and Garkovich 1994).

Symbolic landscapes, or simply landscapes, as Greider and Garkovich refer to them, usually disclose common themes both within and among different narratives. To sum up the point about narratives and identity, self-definitions influence narratives by guiding the points we aim to illustrate in telling our stories (Shanahan 1999; Greider and Garkovich 1994). Thus, focusing on how people tell stories or construct narratives reveals what is important to them and why it is so.

Self-definitions, like that of occupation and community member, shape how we understand our environment (Greider and Garkovich 1994). In this study, the concept of (symbolic) landscapes became a tool for discerning the complex relationships within narratives. While quantitative measures could be used to obtain the significance people confer on place, here it is suggested that qualitative interviewing was more adequate.

THE SAMPLE OF COASTAL COMMUNITIES

The communities that were chosen for study were done so for the Coastal Communities Project, directed by Dr. Jenkins and Dr. Laska of the Center for Hazards Assessment, Response and Technology (CHART) of the sociology department at the University of New Orleans. This study differed from the larger project in one major way—this research focused specifically on residents' understanding of coastal land loss, which was but one element of the larger, more ethnographic whole of the Coastal Communities Project conducted by CHART. That project took a more ethnographic focus within the communities where, in addition to land loss, CHART's research looked more at the totality of respondents' lives, from community and family history to present lives and culture. My focus took only the interpretive experience of land loss and the subsequent impact on identity as its primary focal point.

Whereas ethnography and phenomenology are similar methodological approaches, the one that best fit my inquiry, phenomenology, sought to illuminate the meaning of a lived experience (van Manen 2002). While ethnography is about the *circumstances* of individuals (Hammersly and Atkinson 1995), phenomenology focuses on the meaning which individuals give to their experiences (van Manen 2002; Creswell 1997). In other words, ethnography is about people, and phenomenology is about the experience. Thus, for this study, the analysis took shape through an attempt to uncover what it was like to experience the particular phenomena of coastal land loss in Louisiana.

Although sampling procedures and data collection were the same for both projects, the data set for this research—residents' discussion of coastal land loss—was only a part of all the data for the Coastal Communities Project. The Coastal Communities Project's data set included much more personal, familial, and place history along with much richer descriptions of the circumstances of respondents' lives, which is indicative of a more ethnographic approach.

Six parishes were chosen for the Coastal Communities Project. Either one coastal community or a small cluster of communities, as directed by Jenkins and Laska, within each parish was chosen as the source for a sample. In some instances, we chose a small cluster over just one community in order to accomplish a more robust sample from that parish. Also, linear community development along bayous and rivers made the selection of a cluster of communities practical in three of the parishes—Terrebonne, Plaquemines, and St. Bernard.

Many of coastal Louisiana's rural communities have developed in an interdependent, linear fashion, displaying a mix of activity from fisheries to agriculture to oil and gas extraction and related industrial activity (Gramling and Hagelman 2004). For example, oil and gas may dominate one community, while manufacturing parts for oil and gas extraction may stand out in the neighboring community. Supplementing the industrial activity, fishing, which supplies seafood to the other communities, may be primary to the next community. For the sake of greater reliability, linear development of this nature made sampling residents from a cluster of communities, as opposed to just one, necessary.

A degree of dependence between towns is also important to the maintenance of a sense of community for these areas. In St. Bernard, Terrebonne, and Plaquemines, this linear development helps to "maintain a distinct identity, fostered by kinship and friendship networks" (Gramling and Hagelman 2004, p. 17). Considering the overall coast and the amount of time allotted for data collection (one to one and a half years), Laska and the team decided that sampling from six parishes was attainable and would also provide a comprehensive picture of Louisiana coastal residents who face land loss.

Collaborative meetings among a variety of scientists played a large role in community selection. Coastal geologists, a coastal geographer, two political scientists, two sociologists, and an urban scholar all played a role in choosing communities for study. The now departed Dr. Shea Penland, a coastal geologist, tenacious advocate for coastal restoration and director of the Pontchartrain Institute for Environmental Sciences at the University of New Orleans, and Dr. Shirley Laska, a sociologist, director of CHART, and the initiator of this project (also at the University of New Orleans) were invaluable drivers of this process due to their work in this area over the past few decades.

Our first goal was to identify parishes as coastal. Variables were established for how coastal parishes were to be defined and considered for study. We used the "coastal zone boundary" as defined by the Louisiana Department of Natural Resources as a guiding framework from which to work. The primary variables for parish selection were physical factors (significant land loss), coastal related economic activities (oil and gas activity, fishing industry), and social/cultural (coastal occupations and recreational uses of the coast), all of which helped us get a sense of the importance of the coast to the communities' populations. We then analyzed various data sources for Louisiana parishes. We looked at census data and other Louisiana data sources, including data from the Louisiana Department of Wildlife and Fisheries and the Louisiana Department of Natural Resources. We reviewed data on population, occupation, fishing licenses (commercial and recreation—both freshwater and saltwater distributions), and ethnicity/race. Our goal was to achieve a sample that was distributive across the coast. Even though much of Louisiana's coast is, in general, culturally and demographically homogeneous, we obtained a sample that broadly represented the differences across the southeastern coast of Louisiana while using one central coastal community for comparison.

Based on the criteria described above, the team chose six locations that had a prominent relationship with the coast. The parishes the team chose to locate the study within are Jefferson (specifically the community of Grand Isle[2]), Terrebonne, Plaquemines, St. Bernard, and Orleans, all in the southeastern region of the state. We also conducted research in the community of Delcambre that sits on the border of Iberia and St. Mary's Parish in the southcentral region of the state. The southeastern region experiences significantly more land loss than any other region of the state, and this led to some regional differences in narratives which were discussed in chapters four and five of the text.

In choosing communities within parishes for study, it was advised by Jenkins and Laska that we analyze block data from the chosen parishes and review data just as was done for parish selection but with added variables. These additional variables included local land use (i.e., residential, commercial, zoning, protected areas, private and public fishing), community infrastructure (school, places of religious worship, city government buildings), geographical maps and tables (total land and water area[3]), social characteristics (language, marital status, education), economic characteristics (labor force, place of work, occupation, income), physical characteristics of housing (number of rooms, telephone, vehicles, farm), and financial characteristics of housing (value, rent).[4] Based on observed differences in this data, we chose communities and then conducted searches of newspaper articles to get an idea of current and past issues within each community. Prior to entering a community, we conducted a secondary analysis (academic, historical, and mass media). This analysis solidified the choices of communities from which to sample, and gaining access to the communities was then begun.

SNOWBALL SAMPLING AND COMMUNITY RESIDENTS

Dr. Pamela Jenkins of the University of New Orleans, a community sociologist and director of the Coastal Communities Project, accessed informants, established initial contacts, and led the research team in gaining entrée into the different communities. Informants provided us with the names and phone numbers of residents (Singleton and Straits 1999, pp. 339, 348) and, in some instances, researchers involved in the project acted as informants and provided contacts.[5] In other instances, links to other researchers in CHART, the Pontchartrain Institute, area universities, civic personnel, and other acquaintances served as informants and provided resident contacts in communities.

We established a list of contacts for each community and, from there, compiled lists of possible interviewees. Also, in the case of Grand Isle and Terrebonne, Jenkins and Laska conducted informant interviews with community contacts. Informant interviews of contacts provided further entry into communities and served to gain specific "on the ground" information about the communities (Singleton and Straits 1999, pp. 339, 348). These initial interviews preceded fieldwork in a community. In cases where contacts were interviewed prior to the entering of a community for informational purposes, they were reinterviewed during data gathering and included as part of the data set. In order to preserve the integrity of establishing the salience of coastal land loss to place attachment, we told community informants and contacts only that we were researchers wishing to study life in coastal towns.

In marking the building of a snowball sample for interviews, residents' names were obtained, they were contacted, and interviews were arranged.[6] Using a process of chain referral, we asked them to provide names and phone numbers of other members of the target population, whom we then contacted, interviewed, and solicited more names from, and so on (Singleton and Straits 1999). In this way, we established a range of contacts in each community which provided us with a more robust sample that would "represent a range of characteristics in the target population" (Singleton and Straits 1999, p. 163).

Although snowball sampling carries the danger of a homogeneous sample, we tried to achieve class variation within our sampling, an effort which resulted in a sampling population that ranged from poor to upper-middle-class residents among and within communities. We had less success achieving variation by race. Many of the communities are mostly white, and entry into nonwhite areas was difficult. However, we did achieve some racial variation, as discussed in more detail below in the "Data" section which outlines the progression of fieldwork in the various communities.

DATA

The unit of analysis was the individual resident of a Louisiana coastal community. My focus was on seeking to understand how an independent variable (coastal land loss) impacts respondents' attachment to place. Part of the data-gathering process, as determined by Jenkins, was aimed at gaining insight into the individual's environment—their place; thus, studying the social history of place became an integral part of the research procedure. We conducted extensive historical studies of each of the research communities, providing ourselves with an analytical context for conducting interviews. A synopsis of each historical study was provided in chapter three.

Researching the history of the communities set the stage for 126 interviews with a total of 141 respondents[7] that were collected through snowball sampling. The sample consisted of 82 males and 59 females. Historically, many ethnicities have contributed to this region; however, the communities that the sample was drawn from were largely self-described as white (U.S. Census 2000).[8] As a result, 121 out of the 141 respondents were white. South Terrebonne Parish had a significant Native American population, and 6 Native American respondents were obtained there for a total of 8 Native American respondents overall. Plaquemines Parish had a significant Asian/Asian American population where 4 respondents were obtained producing 5 Asian/Asian American respondents overall. Plaquemines also produced 5 African American respondents for a total of 7 overall. The socioeconomic makeup of the communities was largely working to middle class, and while there was a broad representation in socioeconomic status among respondents, the majority of respondents reflected this demographic.

EXPERIENCES, THE INTERVIEW GUIDE, AND CATEGORIES FOR ANALYSIS

The interview guide prompted the respondent to build a narrative of place through personal history. Narratives were constructed around experiences of place, and this produced categories for building an analytical framework—for example, family, work, childhood, experience with storms, and land loss, the category of primary concern for this research. Categories were established and continually reformulated until all data were incorporated. Specifically, place attachment theory (Altman and Low 1992) and the concept of symbolic landscapes (Greider and Garkovich 1994) informed interview questions which shaped the major topics respondents talked about, except for that of land loss, which no interview question addressed. The theories and subsequent interview questions then constituted analytic categories that were revised in analysis until all the data fit into the categories.

So categories, for the most part, came from responses to interview questions and these came from theory. Accrued biographical experiences that account for place attachment and the self-definitions that account for symbolic landscapes led to questions about personal history and place. Greider and Garkovich's concept of (symbolic) landscapes states that when change to the environment occurs, people negotiate the meaning of that change, and this idea led to questions about change in the interview guide. The categories that resulted from the actual interview process are referred to here as themes. Again, the focus of this study was on the analysis of residents' understanding of their experience of coastal land loss and what that means for their place attachment constructs. So, for instance, narration by residents about land loss was categorized as the theme *land loss*. All land loss data was extracted from the larger interviews, and upon analysis of those extractions subthemes emerged from within the primary theme of land loss (Creswell 1997). Subthemes were named in accordance with the different meanings respondents conferred upon the experience of land loss and so reflect the analytical framework.

ANALYTICAL FRAMEWORK

I incorporated elements of *interpretive phenomenological analysis* (IPA) as outlined by Smith (2004) into the analytical framework. IPA is the science of meaning and meaning-making, and this reflects the idea that our experiences exist for us through the meanings we give them (Smith 2004). When we interpret an experience, that interpretation is meaning-making, and meaning-making is a transformation of the experience. It is not the experience itself; whenever we reflect upon an experience, it is transformed in some way (Smith 2004). Narrative production through interviews is exactly this (van Manen 2002): an embodied *meaning*, what this study attempted to capture—how subjects constructed meaning from their experiences by developing a narrative.

Nevertheless, while the procedure of IPA is concerned with the perceptions of individuals, it also acknowledges the role of the analyst in interpreting meaning residents have given to their experiences (Smith 2004): "The participant is trying to make sense of their personal and social world; the researcher is trying to make sense of the participant trying to make sense of their personal and social world" (Smith 2004). With this process taken into consideration, IPA consists of three basic elements—inductive, idiographic, and interrogative (Smith 2004). The inductive and idiographic were employed in the in-depth analysis which makes up the body of this book, and the concluding chapter touches on the interrogative.

Interpretive phenomenological analysis is inductive in that it allows for unexpected themes to arise, and the data collection method reflects induction because it elicited a narrative about place that did not address coastal land loss. However, telling a story about place led respondents to talk about what was important to them about place, and, as it turned out, for most this involved addressing land loss (Smith 2004).

Thus, looking at the primary theme of land loss as a whole in relation to a narrative of place attachment constituted the first phase of analysis. What resulted was a general picture (presented in table form above) of where coastal land loss was brought up in a narrative, its context, and whether it was a "running theme" throughout the interview.[9] Induction occurred again in the idiographic phase as subthemes emerged from the primary theme of land loss (Smith 2004).

In the idiographic phase of analysis, the theme of land loss was examined for each case (in each interview), reaching a point of closure about each unit before moving to the next. This is usually done with a very few subjects, but since this sample is rather large, a general examination and point of closure was reached on resident discussions of land loss (Smith 2004). Using Atlas.ti analysis software supplied by CHART, the summary examination and point of closure was connected to the land loss theme for each respondent.

After this stage of analysis, I went back into the interviews focusing solely on the primary theme of land loss. I then separated statements of meaning and grouped them according to commonality. Creswell (1997) calls these meaningful statements by respondents "meaning units." In these "meaning units," it was found that coastal land loss was illuminated by the concept of symbolic landscapes—the symbolic meaning of land loss. Grouping statements about land loss according to common elements of meaning created subthemes.

Inductively discovered subthemes were described in detail in chapter four, on residents' attachment to place, and in each of the themes presented in chapter five where meaning was extracted and described in an interpretive phase of analysis. In a general way, an essential and common meaning was presented at the end of each theme and then in the conclusion of chapter five. The reader should come away with the "essential, invariant structure" of experiencing coastal land loss—in short, what it is like to live with this disaster (Creswell 1997; Smith 2004).

Gaining this understanding holds value: moving closer to the particular brings us closer to the universal (Smith 2004). By delving into what the experience of coastal land loss means for residents, we can understand how we are similar to people who we believe live in very different circumstances from our own. Experiencing coastal land loss also carries indications for understanding the meanings many brought into Hurricanes Katrina and Rita. Understanding how land loss is experienced provides a context for Hurricanes Katrina and Rita and, universally, how we might deal with corresponding situations (Smith 2004).

After this detailed interpretive phase of analysis, the concluding chapter moved into the interrogative. The chapter outlines what the findings mean for current coastal restoration policy, especially in light of the hurricanes of late summer 2005, as well as more general policy where place attachment is part of the decision-making process.

Policy makers need to know more about how people think, especially in the realm of disasters. In this way, the qualitative phenomenological approach employed here is useful in that it allowed residents to develop a composite narrative of place. More than that, the phenomenological instrument developed for this study allowed subjects to broach coastal land loss on their own terms.

VALIDITY AND RELIABILITY

Although the validity and reliability of this project can be discerned from much of the above discussion, some clarification is needed. In qualitative research, validity pertains to whether or not the research being conducted accurately measures what it is claiming to measure. Reliability refers to how consistently instances such as interview statements are assigned to the same category by different analysts or by the same analyst at different times (Hammersley and Atkinson 1992). But first, validity is considered.

A qualitative phenomenological method was the most appropriate way to determine how respondents understood coastal land loss. Knowing how respondents subjectively understand the phenomenon using in-depth interviewing does this better than survey methodologies. Again, during community/interviewee research, residents believed the researcher was only interested in their lives as Louisiana coastal residents. This allowed residents to broach land loss on their terms and to develop their own relevance. In this way, we compensated for "social desirability," a common validity problem in qualitative research where responses may be an element of what the respondent believes the researcher wants.

Also, the number of parishes chosen to sample from and the number of interviews, 126, provided a broad and valid sample of coastal residents who were exposed to land loss. Snowball sampling was used, beginning with several different contacts who provided more resident contacts. Interviews with residents who ranged in sociodemographics were purposely sought out. Seeking an array of residents further ensured the validity of the sample and also increased reliability during analysis.

Creswell (1997) suggests that qualitative research should prove reliability by determining the accuracy of analysis and findings. For this analysis, accuracy was achieved by corroborating through intercoder reliability. A colleague, Traber Davis, who was also an interviewer for this project, coded a series of interviews. Atlas.ti, a qualitative analysis software program, was used, and Davis and I employed the same set of analytic codes. I compared our coding of identical interviews and then Davis and I discussed and reworked discrepancies until intercoder reliability was achieved.

A SEVENTH THEME

Originally, there was a small amount of data that constituted a seventh theme. The theme encompassed neutral comments by respondents and was simply called *change*. The comments were significantly fewer and far less voluminous as compared with the other themes. Additionally, the passages in the *change* theme did not represent residents who held unbiased definitions of land loss. In fact, not one respondent who mentioned land loss expressed no opinion on the matter. There were simply some statements within narratives that had an air of neutrality and ambivalence and only indicated the changes brought by land loss. These passages lacked specific political, philosophical, or emotional overtones.

An example of a statement that was coded as *change* came from Mitchell (LC), who said, "Right up there where the little island used to be at the mouth of the Chef's Pass. I remember fifteen years ago that was an island." Mitchell's comments are passive. Although he is actively remembering the condition of an area, he is simply recalling what was once there. So, in passages such as these, it appeared that the narrator's intention was unclear, ambivalent and only that *change* had occurred. This theme was not given separate coverage here due to the fact that it did not represent residents' full intentions about their experience of land loss. However, there were enough statements that simply conveyed that change was occurring that the small theme warrants mentioning.

NOTES

CHAPTER 1

1. The Center for Hazards Assessment, Response and Technology at the University of New Orleans granted permission to use the data for this book. Original research for this book was supported through an EDI/Special Project Grant (B-01-SP-LA-029) from the U.S. Department of Housing and Urban Development (HUD) and the Center for Hazards Assessment, Response and Technology at UNO, LA.
2. Meanings of coastal land loss are outlined and explored in depth in chapters four and five.
3. Dr. JoAnne DeRouen Darlington of the University of Louisiana at Lafayette led most of the interviews in the first community of Grand Isle while I trained and got used to the methodological style of open-ended interviewing. Dr. DeRouen Darlington and I conducted many of the Grand Isle interviews together.
4. Also see Pred 1983; Relph 1976; Tuan 1979.
5. Chapters four and five explore how the meanings that the residents give to coastal land loss are reflective of identity.

CHAPTER 2

1. In 1950 Louisiana held 40 percent of the nation's wetlands. Substantial land loss has led to the current level of 30 percent (LDNR 2004). It also should be noted that these estimates are prior to Hurricanes Katrina and Rita in late summer 2005. While it is doubtful that the storms significantly affected these bottom-line estimates, significant land loss was incurred.
2. The current rate of loss is at about twenty square miles per year; however, this is due to the decreasing availability of wetlands to lose.
3. This wait time for some regeneration has been updated to "the next several growing seasons" by the USGS-maintained LACoast.gov Web site.
4. These projects continue to meet with resistance. In October of 2006 the Coastal Wetlands Planning, Protection and Restoration Act Task Force—for the second time in a year—rejected continued funding for a project to divert Mississippi River water into Bayou Lafourche and build wetlands. Funding was rejected under criticism from representatives of the U.S. Army Corps of Engineers (Wold 10/19/06).
5. CWPPRA was supplemented by a "no net loss of wetlands" rule by the first Bush administration which required those who destroyed wetlands through development to mitigate that loss in another area (van Heerden 2006). Industry has found various ways around this rule.

CHAPTER 3

1. The social history of the coastal communities presented here is mostly since European settlement. This is not meant to further decenter native communities that occupied the region before the arrival of Europeans. The reasons for my concentration on recent history are due to the limited knowledge in the area, the focus of this book, and the fact that most residents in the region today descend from European settlers, some native peoples, and some communities of the Carribbean and South America, all of which are discussed.
2. The historical outlines here are not meant to provide an explanation for the development of residents' attachment or their experiences of land loss. For one thing, this historical discussion is far too brief and basic in that it provides only general economic, political, and social shifts over the past two centuries or so. A more detailed and nuanced historical account of southeastern Louisiana and the communities under study is beyond the focus of this book and has been given by others, some of which are referenced below. The historical treatment presented here is intended only to serve as a sort of contextual scaffolding where community members' discussions of their experience of land loss and their attachment to place can be generally situated.
3. Choosing this geographical cluster of communities provided a more robust representation of the coastal area of the parish and a more precise picture of coastal Louisiana residents than selecting just one community within the parish to gather data.
4. Poststudy, it was thought by the research team of the larger Coastal Communities Project by CHART that Delcambre was not essential to the aims of the study. Such are the trials of social science. However, the narratives of the community are relevant and add some interesting insights to the data set. Thus, I have kept the interviews from Delcambre in this analysis.

CHAPTER 4

1. Also see Pred 1983; Relph 1976; Tuan 1979.
2. Also see Breakwell 2000.
3. There are historical and biological components to the social construction of places. For a fuller explanation of these processes, see Tuan, *Topophilia* (1974).
4. The place attachment process does not follow such rigid consecutive steps; however, the explanation of the process in this way is useful for clarity and understanding.
5. It should be obvious that identity is also shaped by social contexts.
6. There were 126 interviews, but 141 respondents. Some interviews consisted of two residents being interviewed together; however, these were counted as one interview.
7. This last part of the interview, in which respondents were asked to discuss changes to place, is the closest the interview guide came to addressing land loss.
8. The "life world" is reality as it exists through our perceptions and experiences.
9. Furthermore, narratives not only reflect identity, but they consist of general commonalities based on a shared culture, geographic locale, cultural definitions of class, race, age, gender, and occupation, as well as the particular historical moment (Cantrill 1998). When we relay a narrative, it is an experience that we believe can be shared, so as Denzin further notes, narratives do not necessarily position the narrator, or more specifically the self, at the center of the story (p. 44). We want the listener to empathize with the point of our story. Consequently, narratives are often based on "anecdotal, everyday, commonplace experiences" (Denzin 1989, p. 44).
10. Take note that in the residents' interview passages the names of the respondents (pseudonyms) will be followed by an abbreviation of their community in parentheses, and then their age and

an indication of their occupation. Respondents from Grand Isle will be identified as (GI), Terrebonne as (T), Plaquemines as (P), Delcambre as (D), St. Bernard as (SB), and Lake Catherine as (LC). Although presentations of residents' narratives are often presented together, they were interviewed separately unless otherwise indicated.
11. Many scholars have noted that the more time we spend in or with places, the more likely we are to become attached and thus identify with place (Tuan 1974; Greider and Garkovich 1994; Altman and Low 1992; Proshansky et al. 1983).
12. Altman and Low (1992) illustrate how place attachment forms in this way.
13. Soren and Joseline are married, but they were interviewed separately.
14. This is consistent with other research. See Clayton and Opotow (2003).
15. See Tuan (1974).
16. See Greider and Garkovich (1993) for an explanation of (symbolic) landscapes and change.
17. Also see Peacock and Ragsdale (1997).
18. This places the scientists and engineers as outsiders who spend time indoors as opposed to locals who have more contact with the land and thus are more dark complected.
19. Kaplan and Kaplan 1989; Herzog et al. 1997; Korpela et al. 2001.
20. Kroll-Smith and Couch (1993) reiterate this, stating that there is "sufficient empirical evidence" that a change to the environment produces a change "in the organization of personal qualities each of us attributes to ourselves" (p.55).
21. Indeed, this foreshadowing came to fruition with Hurricanes Katrina and Rita. But even before these devastating storms, respondents expressed a postdisaster acknowledgment of attachment.
22. The first part of Edmund's response is to the question "What place or places are important to you?" and the second part is in response to "What would you say is good about it and what needs to be changed?"
23. Carmen's passage is in response to "What is something that you have learned in your life that has stayed with you?"

CHAPTER 5

1. "I Dream a Highway," written by Gillian Welch and David Rawlings. Published by Irving Music Inc./Say Uncle Music/Cracklin' Music (adm. by Bug) BMI.
2. Residents chose to talk about land loss in the context of their attachment to place most frequently (presented in chapter four). Subsequently, the themes presented in this chapter follow from the most frequently occurring to the least. These themes encompass all of the residents' discussion of land loss except for a small portion of their narratives on the issue which they framed in a neutral way and which simply noted the change to the physical landscape that land loss had brought.
3. For an illuminating full-range account of the role trees play in our lives—historically, chemically, psychologically, and otherwise—see Nalini Nadkarni (she is a tree canopy biologist), *Between Earth and Sky: Our Intimate Connections to Trees* (2008). For a more concise account of cultural identity and trees, see Robert Somner (2003), "Trees and Human Identity" in Clayton and Opotow, eds. (2003).
4. Many trees may have already begun dying from saltwater intrusion during the childhood of many residents but were not noticeable at the time.
5. The knowledge of what might be lost was given more urgency and nearly realized with Hurricanes Katrina and Rita and, more recently, Gustav and Ike in 2008. Remember that these narratives were given two to three years before Hurricanes Katrina and Rita. Even before these storms, however, residents were well aware of potential disaster.

6. Cheyenne's passage was also coded for the *attachment* chapter. However, it seems clear that the primary intent of her passage is to convey the *damaging consequences* of land loss over her making a point about the degree of her attachment to place. The thread of damage that she weaves through her passage makes clear her intent. This thread is evident in statements like "if it wasn't for the cement road, my yard would be marsh" (in the original passage), "because that's water right there. That used to be land," and solidifying her intent by summing her passage with "'That's all the proof I need that fifty acres a year are going off in erosion."
7. Mitchell (LC), fifty-two-year-old shipfitter: [Interviewer] Mitchell, explain wash out. [Mitchell] You order a load of concrete. The truck comes out, he dumps the concrete and forms it up. When that truck goes back, you've got to wash that concrete out of that big container. What they'll do is they'll hose it all out of there and then dump that out on the ground. It doesn't have as much to guard the line then so it's not true concrete then. What happens is it dries up and hardens up on the ground. And then they eat it up with a big piece of machinery, scoop it up and throw it in the back of the truck. Well, it's almost like a concrete. It gets hard after a long time. And it's a pretty good stabilizer.
8. In pure volume, residents expressed their attachment to place (covered in chapter four), the damaging consequences of the loss, and restoration issues the most. For each of the next three themes, there were about half as many statements as each of the previous three. However, despite numbers of statements, the passages within these themes conveyed the profound meaning of experience no less than the others, but there were fewer of them.
9. During this passage the interviewer asked Phyllis to clarify her statements. The breaks in this passage indicate where this occurred.
10. Sven of Delcambre is coded as not having talked about land loss because he brought up the issue after being asked by the researcher about the event. I have not included passages of residents for whom this occurred; however, I have included Sven's words here due to the relevance of comparison with those of Terrebonne residents.
11. The residents of Delcambre may be only beginning to experience what residents of more southeastern parishes are feeling. And now, after they suffered significantly from Hurricanes Katrina, Rita, and, most recently, Ike, realizations of land loss may have come to the people of Delcambre a lot more quickly than they might have expected.
12. Thus, there is impetus for future analysis of heightened awareness of place attachment within a slow onset disaster compared with regions and communities where significant place attachment is expected but is relatively unthreatened.
13. And as we now know, Hurricanes Katrina, Rita, and, most recently, Gustav and Ike exploited the loss of protective coastal land and gave credence to the admonitions that Jenny and others sent out.
14. Albert was then asked, "How has that affected you guys as far as your ventures and stuff?"
15. Many of the passages on individual and community agency came from Lake Catherine. The geography of the community lent itself to individual mitigation more than most other communities. However, while residents of Lake Catherine spoke of their own efforts, it was narrow in the sense that their efforts were only on their property and didn't take into account the land loss occurring just yards away from their homes. In this way, their accounts represent a sort of closing of the ranks from the larger disaster that encircles them.
16. Of course, from this analysis of one-time interviews, we can't be certain that a continuous event like Louisiana's land loss causes a more constant heightened anxiety. To be sure, there must be more time-oriented studies, preferably with the use of community-based ethnomethodologies.

That being said, the extension of data collection (six parishes over nearly fifteen months) and the words of these residents do indicate the likelihood of a continuously sharp anxiety among community members. This more constant awareness produced the anxiety, uncertainty, and desperation expressive of a fragile identity.

CHAPTER 6

1. In a way, this process is similar to how many of us have become estranged from community. Conversely, this alienation from community has led to initiatives to re-create community, such as neighborhood and park cleanups, neighborhood farmers' markets, and the like.
2. As I first learned from my father-in-law, who owns a small farm in central Georgia, clearing dense undergrowth is sometimes necessary because primary, and oftentimes secondary, forest growth was clear-cut; thus there is no tree canopy to regulate the undergrowth. As a result, this undergrowth grows so densely that it strangles the diversity indicative of a healthy locale.
3. Daniel Jaffe's 2007 *Brewing Justice: Fair Trade Coffee, Sustainability, and Survival* provides an extensive overview of the ascendance of the Fair Trade system, its benefits and downsides for growers in Mexico, and its connection to the global economy. Jaffe also explores the challenges that Fair Trade growers face, such as stagnating prices, as they have gained more market share with purchases by U.S. corporations. He also explores the struggle and rift within the movement of whether to hold on to the principles of their sustainable system and remain small or go for more market share but sacrifice some of those principles.
4. In *Radical Ecology: The Search for a Livable World* (2005) Carolyn Merchant outlines the philosophical movement of Deep Ecology and others such as postclassical sciences, feminism, and reconstructive knowledge as providing the framework for the possibility of living out a new, ecologically sustainable worldview in the twenty-first century. The principles of this worldview include "rejecting the man-in-environment image in favor of the relational, total field image," fighting against resource depletion, and "local autonomy and decentralization" (p. 92). Like the attachment of coastal residents, this involves establishing a view of nature that is part of a society's self-interpretation where the affinity and emotional attachments for a local landscape are the "prime motivators for an ethic of care" (p.111).
5. An important addition to this equation is a study released in June of 2009 in the journal *Nature Geoscience* by geologists Michael Blum and Harry Roberts stating that even under the best of circumstances and the building of large, industrial engineering projects, the state will lose four thousand to five thousand square miles of coastal land. This might not be all bad, however. A crack in industrial restoration is exposed here.
6. Originally stated in Fear and Edwards 1995.

APPENDIX

1. While this finding is relevant, it is slightly limited because in 3 of the 36 narratives, I or other interviewers were probing a resident. If a resident broached an issue such as land loss while discussing another topic, interviewers would wait to probe so as not to disrupt the flow and construction of the resident's narrative. Thus, an interviewee may have introduced land loss in the first phase of their narrative; however, the interviewer may not have had an opportunity to probe until the next portion of the interview. After searching each interview that constituted land loss as a running theme, 3 of the 36 were found to be probed by the interviewee in subsequent portions of

the interview (second and/or last third). These consisted of one interview of a couple from Grand Isle, one from Terrebonne, and one in Lake Catherine. That being said, these respondents talked about the issue at length and returned to the issue after moving into other areas of their narrative.
2. We didn't necessarily choose Jefferson Parish so much as we chose the incorporated community of Grand Isle for its unique relationship to Louisiana and the coast.
3. Communities that are somewhat inland were not excluded from selection and, in fact, were equally viable with those that sit more directly on the coast due to the nature of southeastern Louisiana's land mass. The area is comprised of wetlands that have many bayous and inlets which would make it likely for residents to view themselves as "living on the coast" despite not being literally situated on the Gulf of Mexico.
4. We did not have strict predefined criteria for community selection but looked at these characteristics to attain community variation and similarity.
5. Contacts were residents provided through informants who had a broad knowledge of the community and its residents.
6. I was the lead field researcher, being advised by Jenkins, and conducted most of the interviews; however, three other researchers were part of the interview team (see acknowledgments), and as a whole they conducted about 30 percent of interviews.
7. Some interview sessions consisted of more than one respondent but are counted as a single interview.
8. The parishes chosen showed more ethnic diversity than was gathered in the sample; however, I focused on "coastal" communities. Many of these parishes are largely urban. The communities I focused on were in the outlying areas of the parish and areas that residents perceived as being "on the coast." These communities are largely white. A parish such as Orleans has a large African American population who reside mostly in New Orleans; however, the focus here was on the community of Lake Catherine, which is described by residents as an "inner coastal community." The community is almost entirely white. The data set of Lake Catherine had one black resident out of nineteen respondents. The rest were white.
9. For this phase of the analysis, a theme was defined as "running" if a respondent broached land loss in each segment of the interview—i.e., during the history of place portion of the interview guide (the first portion), experience with storms (the second portion), and/or change (the last portion). This has limitations because a respondent might talk about land loss a great deal in the first phase of the interview, but not bring the issue up in the remainder of the interview. This is compensated for in the idiographic phase of the analysis.

REFERENCES

Alpert, Bruce. 2007. "Louisiana's Coastal Restoration Plan Wins Approval." *The Times-Picayune*, November 30.
Altman, I., and S. M. Low, eds. 1992. *Place Attachment*. New York: Plenum Press.
American Planning Association. 1997. "Barataria-Terrebonne Estuary Comprehensive Conservation and Management Plan—American Planning Association 1997 Planning Awards." *Planning* 63 (4): 1–5.
America's WETLAND. 2005. *America's Wetland: Campaign to Save Coastal Louisiana*. http://www.americaswetland.com/.
America's WETLAND. 2006. *America's Wetland: Campaign to Save Coastal Louisiana*. http://www.americaswetland.com/.
Ancelet, B. J. 1989. "Cajun Land." *Southern Exposure* 17 (3): 52–53.
Austin, Maureen E., and Rachel Kaplan. 2003. "Identity, Involvement and Expertise in the Inner City: Some Benefits of Tree-Planting Projects." In S. Clayton and S. Opotow, eds., *Identity and the Natural Environment: The Psychological Significance of Nature*. Cambridge, MA: MIT Press.
Barras, J. A., S. Beville, D. Britsch, S. Hartley, S. Hawes, J. Johnston, P. Kemp, Q. Kinler, A. Martucci, J. Porthouse, D. Reed, K. Roy, S. Sapkota, and J. Suhayda. 2003. *Historical and Projected Coastal Louisiana Land Changes: 1978–2050*. USGS Open File Report 03-334.
Barry, John. 1998. *Rising Tide: The Great Mississippi Flood of 1927 and How It Changed America*. New York: Simon and Schuster.
Boyer, M. C. 1994. *The City of Collective Memory: Its Historical Imagery and Architectural Entertainments*. Cambridge, MA: MIT Press.
Breakwell, G. M. 2000. "Social Representational Constraints upon Identity Processes." In K. Deaux and G. Philogene, eds., *Representations of the Social*. Oxford: Blackwell Publishers.
Brown, B. B., and P. B. Perkins. 1992. "Disruption in Place Attachment." In I. Altman and S. M. Low, eds., *Place Attachment: Human Behavior and the Environment*. New York: Plenum Press.
Brown, Matthew. 2005. "Plaquemines Oyster Farmer Suit Dead; Nation's Highest Court Declines to Hear Case." *The Times-Picayune*, May 24.
Bruckner, Monica. 2008. "The Gulf of Mexico Dead Zone." Microbial Life Educational Resources. http://serc.carleton.edu/microbelife/topics/deadzone/#cause.
Burley, David, Steve Couch, and Steven Kroll-Smith. (forthcoming). "Commonsense, Science and Participatory Research: A Case for Colloquial and Scientific Knowledge in Environmental Controversies." Contact authors for more information.
Burley, David, Pamela Jenkins, Shirley Laska, and Traber Davis. 2007. "Place Attachment and Environmental Change in Coastal Louisiana." *Organization and Environment* 20 (3).

Clayton, Susan. 2003. "Environmental Identity: A Conceptual and Operational Definition." In S. Clayton and S. Opotow, eds., *Identity and the Natural Environment: The Psychological Significance of Nature*. Cambridge, MA: MIT Press.

Clayton, Susan, and Susan Opotow, eds. 2003. *Identity and the Natural Environment: The Psychological Significance of Nature*. Cambridge, MA: MIT Press.

Creswell, John. 1997. *Qualitative Inquiry and Research Design: Choosing Among Five Traditions*. London: Sage.

Cronon, William. 1983. *Changes in the Land: Indians, Colonists, and the Ecology of New England*. New York: Hill and Wang.

Crumley, C., ed. 1994. *Historical Ecology: Cultural Knowledge and Changing Landscapes*. Santa Fe, NM: School of American Research Press.

Denzin, Norman K. 1989. *Interpretive Biography*. Sage Series in Qualitative Research Methods, Volume 17. London: Sage.

Eggler, Bruce. 2003. "Property Owners May Soon Get Titles; Lake Catherine Land Set to Be Resubdivided." *The Times-Picayune*, Metro, p.1.

Farber, Stephen. 1996. "Welfare Loss of Wetlands Disintegration: A Louisiana Study." *Contemporary Economic Policy* 14 (1): 92–107.

Fear, Kathleen, and Patricia Edwards. 1995. "Building a Democratic Learning Community Within a PDS." *Teaching Education* 7 (2). http://www.teachingeducation.com/vol7–2/72te.htm.

Freudenberg, William R., and Robert Gramling. 1994. *Oil in Troubled Waters: Perceptions, Politics, and the Battle Over Offshore Drilling*. SUNY Series in Environmental Public Policy. New York: State University of New York Press.

Gebhard, U., P. Nevers, and E. Billman-Mahecha. 2003. "Moralizing Trees: Anthropomorphism and Identity in Children's Relationships to Nature." In S. Clayton and S. Opotow, eds., *Identity and the Natural Environment: The Psychological Significance of Nature*. Cambridge, MA: MIT Press.

Gieryn, T. F. 2000. "A Space for Place in Sociology." *Annual Review of Sociology* 26: 463–96.

Gramling, Robert, and Ron Hagelman. 2004. "A Working Coast: People in the Louisiana Wetlands." *Journal of Coastal Research*. ISSN: 0749-0208.

Greider, Thomas, and Lorraine Garkovich. 1994. "Landscapes: The Social Construction of Nature and the Environment." *Rural Sociology* 59 (1): 1–24.

Gubrium, Jaber F., and James A. Holstein. 1997. *The New Language of Qualitative Method*. New York: Oxford University Press.

Hansen, Liane. 2008a. "Rising Sea Levels Threaten Egypt's Ancient Cities." National Public Radio, April 20. http://www.npr.org/templates/story/story.php?storyId=89660898.

———. 2008b. "In Cairo Slum, the Poor Spark Environmental Change." National Public Radio, April 27. http://www.npr.org/templates/story/story.php?storyId=89956754.

Hecht, Jeff. 1990. "The Incredible Shrinking Mississippi Delta." *New Scientist* 14 (April), no. 1712.

Higgs, Eric. 2003. *Nature by Design: People, Natural Process, and Ecological Restoration*. Cambridge, MA: MIT Press.

History and Geneology in Terrebonne Parish, LA. 1997. www.rootsweb.com/~/aterreb/index.htm.

Iberia Parish Tourist Commission. *Iberia Parish, New Iberia, Louisiana*.

Jaffe, Daniel. 2007. *Brewing Justice: Fair Trade Coffee, Sustainability, and Survival*. Berkeley, CA: University of California Press.

Jeansonne, Glen. 1995. *Leander Perez: Boss of the Delta*. Baton Rouge, LA: LSU Press.

Jensen, Lynne. 2003. "Brazilier Tenant Evictions on Hold; Restrainer Order Issued Until Trial." *The Times-Picayune*, February 12, Metro, p. 1.

Joyce, Christopher. 2005. "Restoration of Gulf Coast Wetlands Poses Challenges." National Public Radio, November 10. http://www.npr.org/templates/story/story.php?storyId=5006890.

Kahn, Peter H., Jr. 2002. "Children's Affiliation with Nature: Structure, Development, and the Problem of Environmental Generational Amnesia." In P. H. Kahn, Jr., and S. R. Kellert, eds., *Children and Nature: Psychological, Sociocultural, and Evolutionary Investigations.* Cambridge, MA: MIT Press.

Kaplan, R., and S. Kaplan. 1989. *The Experience of Nature: A Psychological Perspective.* New York: Cambridge University Press.

Korpela, K. M., T. Hartig, F. G. Kaiser, and U. Fuhrer. 2001. "Restorative Experience and Self-Regulation in Favorite Places." *Environment and Behavior* 33: 572–589.

Kroll-Smith, Steve, and Steve R. Couch. 1993. "Symbols, Ecology, and Contamination: Case Studies in the Ecological–Symbolic Approach to Disaster." *Research in Social Problems and Public Policy* 5:47–73.

LCA (Louisiana Coastal Area Study) Fact Sheet. 2004. *Louisiana Coastal Area Study (LCA), Ecosystem Restoration Study.* http://www.lca.gov/lcaFactSheet1103new9-29-04.pdf.

LCA (Louisiana Coastal Area Study). 2004. *Louisiana Coastal Area Study (LCA), Ecosystem Restoration Study,* full report. http://data.lca.gov/Ivan6/main/main_report_all.pdf.

Louisiana Coastal Area (LCA), LA Ecosytem Restoration Study. 2004. U.S. Army Corps of Engineers, New Orleans District.

Louisiana Collection. *Plaquemines: Where Nature Means Progress.* University of New Orleans, Earl K. Long Library. No. F377.P45C52 La.Col.

Louisiana Department of Natural Resources. 2004. *Louisiana Coastal Facts.* Coastal Restoration and Management Division.

Louisiana State Parks. 2008. *Fort Pike State Historic Site.* www.lastateparks.com/fortpike/fortpike.htm.

LSU AgCenter: Research and Extension. 1998. *Louisiana Cooperative Extension Service St. Bernard Parish.* www.agctr.lsu.edu/wwwac/parish/st_bernard/abtstb.htm.

Marx, Karl. 1844. *Economic and Philosophical Manuscripts of 1844, Karl Marx: Estranged Labour.* http://www.marxists.org/archive/marx/works/1844/manuscripts/labour.htm.

Meitrodt, Jeffrey, and Aaron Kuriloff. 2003. "Shell Game." *The Times-Picayune,* May 4, Section A, p.1.

Merchant, Carolyn. 2005. *Radical Ecology: The Search for a Livable World.* 2nd edition. New York: Routledge.

Morton, R. A., G. Tiling, and N. F. Ferina. 2003. "Causes of Hotspot Wetland Loss in the Mississippi Delta Plain." *Environmental Geosciences* 10:71–80.

Morton, R. A., N. A. Buster, and M. D. Krohn. 2002. "Subsurface Controls on Historical Subsidence Rates and Associated Wetland Loss in Southcentral Louisiana." *Gulf Coast Association of Geological Societies Transactions* 52: 767–778.

Nadkarni, Nalini M. 2008. *Between Earth and Sky: Our Intimate Connections to Trees.* Berkeley: University of California Press.

National Oceanic and Atmospheric Administration (NOAA) Coastal Services Center. 2005. *Hurricane Katrina Inundation Analysis.* Available at http://www.csc.noaa.gov/crs/lca/katrina/.

National Public Radio (NPR). 2008. "Terrebonne Parish, LA., Recovers After Gustav," September 3. http://www.npr.org/templates/story/story.php?storyId=94243983.

National Research Council. 1992. *Restoration of Aquatic Ecosystems: Science, Technology, and Public Policy/Committee on Restoration of Aquatic Ecosystems.* Washington D.C.: National Academy of Sciences.

Opotow, Susan, and Amara Brook. 2003. "Identity and Exclusion in Rangeland Conflict." In S. Clayton and S. Opotow, eds., *Identity and the Natural Environment: The Psychological Significance of Nature*. Cambridge, MA: MIT Press.

Peacock, W. G., and A. K. Ragsdale. 1997. "Social Systems, Ecological Networks and Disasters: Toward a Socio-Political Ecology of Disasters." In W. G. Peacock, B. H. Morrow, and J. Gladwin, eds., *Hurricane Andrew: Ethnicity, Gender and the Sociology of Disasters*. New York: Routledge.

Penland, S., L. D. Wayne, L. D. Britsch, and S. J. Williams. 2000. "The Processes of Coastal Land Loss in the Mississippi River Delta Plain." New Orleans, LA: USGS Open File Report 00-0418.

Proshansky, H. M., A. K. Fabian, and R. Kaminoff. 1983. "Place-Identity: Physical World Socialization of the Self." *Journal of Environmental Psychology* 3: 57–83.

Reed, Denise, and Lee Wilson. 2004. "Coast 2050: A New Approach to Restoration of Louisiana Coastal Wetlands." *Physical Geography*. 25 (1):4–21.

Reeves, Sally K. 1985. "The Settlement and Cultural Growth of Grand Isle, Louisiana." In Philip D. Uzee, ed., *The Lafourche Country: The People and the Land*. Lafayette, LA, Center for Louisiana Studies.

Relph, E. 1976. *Place and Placelessness*. London: Pion.

Schleifstein, Mark. 2008. "Brazilier Island Bought by Trust." *The Times-Picayune*, September 25.

———. 2005. "Coastal Projects Ready to Soak Up New Aid." *The Times-Picayune*, July 31.

———. 2004. "Steps Toward Restoration." *The Times-Picayune*, July 18.

Schultz, Bruce, and Angela Simoneaux. 1994. "Suit Challenges Lake Peigneur Dredging." *The Advocate*, August 25, Metro edition, p. 3B1B.

Smith, Jonathan. 2004. "Reflecting on the Development of Interpretative Phenomenological Analysis and Its Contribution to Qualitative Research in Psychology." *Qualitative Research in Psychology* 1:39–54.

Somner, Robert. 2003. "Trees and Human Identity." In S. Clayton and S. Opotow, eds., *Identity and the Natural Environment: The Psychological Significance of Nature*. Cambridge, MA: MIT Press.

St. Bernard Parish Library. *An Overview of St. Bernard Parish, Louisiana*.

Steilow, Frederich Joseph. 1981. "Grand Isle and the New Leisure, 1866–1893." *Louisiana History* 23: 239–257.

Stoecker, Randy. 2005. *Research Methods for Community Change: A Project-Based Approach*. Thousand Oaks, CA: Sage.

———. "Are Academics Irrelevant?: Roles for Scholars in Participatory Research." Presented at the American Sociological Society Annual Meetings, 1997. http://comm-org.utoledo.edu/papers98/pr.htm.

Stoecker, Randy, and Edna Bonacich. 1992. "Why Participatory Research?" *The American Sociologist* 23:5–14.

Stuart, Susan M. 2005. "Lifting Spirits: Creating Gardens in California Domestic Violence Shelters." In Peggy F. Bartlett, ed., *Urban Place: Reconnecting with the Natural World*. Cambridge, MA: MIT Press.

Sustainable South Bronx. 2008. http://www.ssbx.org/. Retrieved November 5.

Thomashow, Cynthia. 2002. "Adolescents and Ecological Identity: Attending to Wild Nature." In Peter H. Kahn, Jr., and Stephen R. Kellert, eds., *Children and Nature: Psychological, Sociocultural, and Evolutionary Investigations*. Cambridge, MA: MIT Press.

Thompson. 2002. "Fighting the Current." *The Times-Picayune*. July 31, Sports, p. 1.

Tuan, Yi-Fu. 1974. *Topophilia: A Study of Environmental Perception, Attitudes and Values*. Englewood Cliffs, NJ: Prentice-Hall.

———. 1977. *Space and Place: The Perspective of Experience.* Minneapolis: University of Minnesota Press.
U.S. Army Corps of Engineers. 2004. LCA Ecosystem Restoration Project. http://www.mvn.usace.army.mil/prj/lca/.
U.S. Census. 2000. http://www.census.gov/.
U.S. Department of the Interior. 1994. *The Impact of Federal Programs on Wetlands, Vol. II.* U.S. Government Printing Office, Washington D.C.
U.S. Geological Survey—The National Wetlands Research Center. 2005. USGS reports preliminary wetland loss estimates for southeastern Louisiana from Hurricanes Katrina and Rita. Press release, November 1. Available online at http://www.nwrc.usgs.gov/releases/pr05_007.htm.
Van Heerden, Ivor, with Mike Bryan. 2006. *The Storm: What Went Wrong and Why During Hurricane Katrina—the Inside Story from One Louisiana Scientist.* New York: Viking Press.
WaterMarks. 2003. Special Issue: "Global Climate Change and Louisiana's Coastal Wetlands." WaterMarks: Louisiana Coastal Wetlands Planning, Protection and Restoration News, February, #22.
WaterMarks. 2008. "Standing Ground Against Advancing Waters: Acre by Acre, CWPPRA Projects Beat Back Coastal Demise." WaterMarks: Louisiana Coastal Wetlands Planning, Protection and Restoration News, December, #39.
Working Group for Post-Hurricane Planning for the Louisiana Coast. 2006. *A New Framework for Planning the Future of Coastal Louisiana after the Hurricanes of 2005.* Preparation for the report conducted by the Integration and Application Network of the University of Maryland Center for Environmental Sciences with support by the Institute of Water Resources of the U.S. Army Corps of Engineers and the National Research Council. The full report is available at http://www.umces.edu/la-restore/.

INDEX

Andrew, Hurricane, 16, 30, 32
Avondale, 74

Barataria Estuary, 74
Barataria Island, 61, 74
Baton Rouge, 17, 69
Bayou DuLarge, 62, 133
Bayou Savage National Wildlife Refuge, 36, 37
Betsy, Hurricane, 6
Bill, Tropical Storm, 50, 64, 65
Brazilier Island, 37
Buras, 34

Caenarvon Freshwater Diversion Project, 71, 72
Center for Hazards Assessment Response and Technology (CHART), xiii, xiv, 8, 142, 143, 144, 146, 149, 153, 154
Chandeleur Islands, 31, 88
Chauvin, 29, 30, 67
Coastal Change Analysis Program (C-CAP), 20
Coastal Communities Project, xiii, xiv, 8, 9, 142, 143, 145, 154
Cocodrie, 29, 30
Cogenerative dialogue, 129
CSX Railroad, 36

Davis Pond Diversion, 22, 23, 74
Delacroix, 31
Delcambre Parish, 7, 35, 36, 95, 96, 136, 138, 145
Dulac, 29, 30, 66, 133
Dulac-Grand Caillou, 133

Empire, 34

Focal practices, 120, 121, 122, 123, 125
Fort McComb, 36
Fort Pike, 36

Grand Isle, ix, xiv, xv, 7, 19, 27, 29, 48, 52, 54, 55, 60, 61, 84, 93, 96, 98, 100, 101, 136, 139, 145, 146, 153, 155, 158
Grand Terre, 74
Gulf of Mexico, x, 4, 5, 17, 18, 20, 31, 32, 33, 36, 50
Gustav, Hurricane, 6, 15, 31, 116, 132

Houma Indian Tribe, xv, 9, 47

Iberia Parish, 27, 35, 145
Ike, Hurricane, 6, 15, 31, 116, 132
Interviewees: Adam (T), 53, 54, 68, 76, 79, 80; Albert (GI), 100, 101, 156; Alfonse (T), 82; Alfonse (GI), 52; Alicia (LC), 43, 85; Alysha (GI), 84; Anastasia (GI), 90; Art (P), 51, 56, 64–66, 93; Becky and William (LC), 63, 86; Bettie (LC), 99, 102; Bubba (D), 90; Carmen (T), 58, 97; Carrie (GI), 92, 93; Cedric (T), 64–67; Celestine (SB), 60; Charlie (LC), 46, 47; Cheyenne (SB), 3, 4, 41, 66, 67; Christian (SB), 68, 69, 88–90; Christopher (LC), 86; Chuck (T), 55, 94, 96; Claude (T), 70; Conrad (D), 70, 71; Dorothy (SB), 62; Duke (SB), 88,

89; Edmund (T), 57; Gerry (T), 80, 81, 87; Hank (SB), 89, 90; Jackie (LC), 78–80, 94, 106; Jared (SB), 43, 44; Jenny (GI), 55, 56, 98, 99, 156; Jeppa (P), 43, 44, 72, 73, 113; JJ (T), 104; Joseline (T), 46, 155; Kyle (T), 62, 63, 79, 80; Larry (GI), 74, 75; Leroy (P), 45; Lester (SB), 44, 45, 52, 76, 94; Liane (T), 50, 54, 96, 99; Lynda (GI), 64, 65; Lysha (D), 86–88; Morris (T), 105; Paul (P), 43, 77, 102, 103, 106; PJ (GI), 70, 71; Phyllis (SB), 84, 85, 103, 104, 156; Rachelle (T), 82, 95, 96, 99; Robert (GI), 73; Rocky (P), 49, 50, 75, 76; Roger (D), 60–62, 87; Saro (SB), 74, 75, 81; Sissy and Albert (GI), 57; Soren (T), 46, 155; Susan (GI), 44, 45, 54, 55, 85, 91, 96, 97; Sven (D), 95, 156; Sylvan (GI), 48, 49, 56, 60, 61, 63, 85; Tara (LC), 67, 73; Theodore (T), 47, 48, 87, 88, 94; Thomas (P), 43; Tina (T), 67, 68, 82, 93, 94, 96; Tyronne (SB), 52, 71–73, 75, 103, 104; Vivian (GI), 48, 49; Walter (T), 69, 98, 99
Isidore, Tropical Storm, 4, 50, 64
Isle de Jean Charles, 9, 47

Jefferson Parish, 22, 27, 35, 145

Katrina, Hurricane, xiv, 5, 6, 7, 10, 11, 15, 16, 20, 21, 23, 27, 29, 30, 31, 32, 33, 34, 35, 37, 48, 56, 69, 87, 110, 111, 115, 117 121, 122, 126, 132, 138, 150
Kenner, 80

Lafitte, 74
Lafourche Parish, 22, 61
Lake Borgne, 32, 36, 63, 86
Lake Catherine, xiv, 7, 9, 10, 36, 37, 121, 136, 139
Lake Catherine Land Corporation, 36
Lake Peigneur, 35
Lake Pontchartrain, 36, 80
Lili, Tropical Storm, 4, 9, 50, 64
Local science, 129
Louisiana Coastal Area Ecosystem Restoration Plan, 16, 21, 22, 23, 56
Louisiana Department of Health and Hospitals (LDHH), 36
Louisiana Department of Natural Resources, 76, 144

Louisiana Department of Wildlife and Fisheries, 76, 144
Louisiana Division of Environmental Quality, 20
Louisiana Universities Marine Consortium (LUMCON), 30

Mississippi Delta, 17
Mississippi River, 5, 17, 22, 33, 53, 72, 83, 90, 92, 103, 122
Mississippi River Gulf Outlet (MRGO), 22, 32, 62, 88, 89, 90
Montegut, 67, 133

National Technical Review Committee, 23
New Orleans, x, xiii, 4, 5, 6, 15, 19, 20, 31, 32, 36, 45, 48, 55, 60, 61, 64, 67, 69, 80, 82, 90, 93, 98, 122

Orleans Parish, 4, 36, 37

Participatory research (PR), xii, 124, 125, 127, 129, 130
Place attachment, 8, 9, 12, 13, 40, 41, 42, 44, 54, 96, 107, 110, 111, 120, 121, 140, 141, 146, 148, 149, 150
Plaquemines Parish, 7, 22, 27, 33, 34, 74, 75, 121, 136, 138, 139, 143, 144, 145
Pointe au Chene, 133
Port Fouchon, 19
Port of New Orleans Dock Board, 32

Regional depressurization, 19
Remington Oil and Gas Company, 37
Rita, Hurricane, 5, 6, 7, 10, 11, 15, 16, 20, 21, 23, 27, 29, 30, 31, 35, 37, 48, 51, 55, 56, 69, 78

Save the Lake Foundation, 80
Shell Oil Company, 118
Slurry, 22, 118
Society for Ecological Restoration (SER), 118
Solar Cities, 123
Spanish Canary Islands, 31
Spoil banks, 18
St. Bernard Parish, 4, 5, 7, 22, 27, 31–34, 63, 72, 75, 88, 89, 90, 103, 113, 121, 136, 139, 143, 144, 145, 147
St. Charles Parish, 22

Sustainable South Bronx, 123, 124
Symbolic landscapes, 40, 41, 42, 49, 51, 61, 81, 85, 90, 92, 98, 104, 105, 108, 109, 122, 139, 140, 141, 148, 149

Terrebonne Parish, xiv, xv, 7, 10, 27, 29, 30, 31, 61, 88, 121, 136, 139, 143, 144, 145, 146, 147
Toca, 31
traditional ecological knowledge (TEK), and wisdom (TEKW), 124
Trust for Public Land, 37

U.S. Army Corps of Engineers, 17, 20, 21, 23, 32, 33, 37, 56, 76, 78, 100, 101, 102, 103, 117, 118, 127
U.S. National Research Council (NRC), 118

Vermillion Parish, 35
Violet Canal, 63, 86

Working Group for Post-Hurricane Planning for the Louisiana Coast, 16, 20

Yscloskey, 31

www.ingramcontent.com/pod-product-compliance
Lightning Source LLC
Chambersburg PA
CBHW030344240426
43661CB00052B/1740